# MIDWEST BEDROCK

**Heartland History**

*Jon K. Lauck*

KEVIN J. KOCH

# MIDWEST BEDROCK

*The Search for Nature's Soul in America's Heartland*

INDIANA UNIVERSITY PRESS

*This book is a publication of*

Indiana University Press
Office of Scholarly Publishing
Herman B Wells Library 350
1320 East 10th Street
Bloomington, Indiana 47405 USA

iupress.org



*Manufactured in the United States of America*

First Printing 2024

Cataloging information is available from the Library of Congress.

978-0-253-06883-5 (hdbk.)
978-0-253-06884-2 (pbk.)
978-0-253-06885-9 (web PDF)

FOR MY STUDENTS

*I have tried to teach you this one thing,*
*that all the world is sacred.*

# Contents

# Acknowledgments

My first and deepest gratitude is to my wife, Dianne, who appears in many of these pages. Who else would have traveled to twelve Midwest states with me, arranged hotel and campsite stays, climbed to overlooks, descended into old copper mines and missile silos, hiked, watched patiently as I took a dip at yet another ice-cold Lake Superior beach, and rode bicycles with me on trails and roadways across the Midwest for three summers straight?

Dianne is my navigator. On the road, she scouts out the route, whether for the long haul across a state or to find an obscure back road leading us to a surprising Midwest treasure that few travelers have seen. She is my guide in most other ways as well, through more than forty years of marriage and through raising three children to adulthood. Dianne guided me through the writing of this work as well, gently letting me know when I took a wrong turn.

Three of my longtime teaching colleagues from Loras College deserve special thanks. Haikuist and poet Bill Pauly, who passed away in February 2021, shared an office with me for the first years of my career. He also shared with me his profound love of teaching, his reverent regard for the natural world, his deep attentiveness to the written word, and his gentle spirit. Andy Auge and I taught together for thirty-five years. He long ago recognized in me something about which I had not yet become aware: my own devotion to the sense of place. Andy traveled with me to Lake Itasca as described in the Minnesota chapter of this book. And my colleague Dana Livingston, from the fields of Nebraska, taught me—a woods-person—how to love the prairie. When Dana eventually moved back home, he accompanied Dianne and me to the Niobrara River and the Nebraska sandhills.

A thank-you to Jon Lauck, coeditor of *North Country* (University of Oklahoma Press, 2023), who encouraged me to expand my Lake Itasca chapter into this book about special Midwest outdoors places.

Loras College deserves my gratitude as well. Dianne and I met at Loras as undergrads before I returned as a professor. In the forty-plus years I have taught at Loras, the administration has been unfailingly supportive of my writing projects. Special thanks to President Jim Collins, Provost Donna Heald, and to the Faculty Development Committee for granting me a sabbatical and some funding to work on this project.

Lastly, a thank-you to my students of forty-plus years, to whom this book is dedicated.

The Minnesota chapter appeared previously as "All One Thing: Seeking the Source of the Mississippi" in *North Country: Essays on the Upper Midwest and Regional Identity,* Eds. Jon K. Lauck and Gleaves Whitney, University of Oklahoma Press, 2023.

Shorter versions of some of these chapters appeared in my Outdoors/Travel column in the *Telegraph Herald* (Dubuque, IA) newspaper.

Map of the Midwest and locational sites drawn by Angela Koch.

# MIDWEST BEDROCK

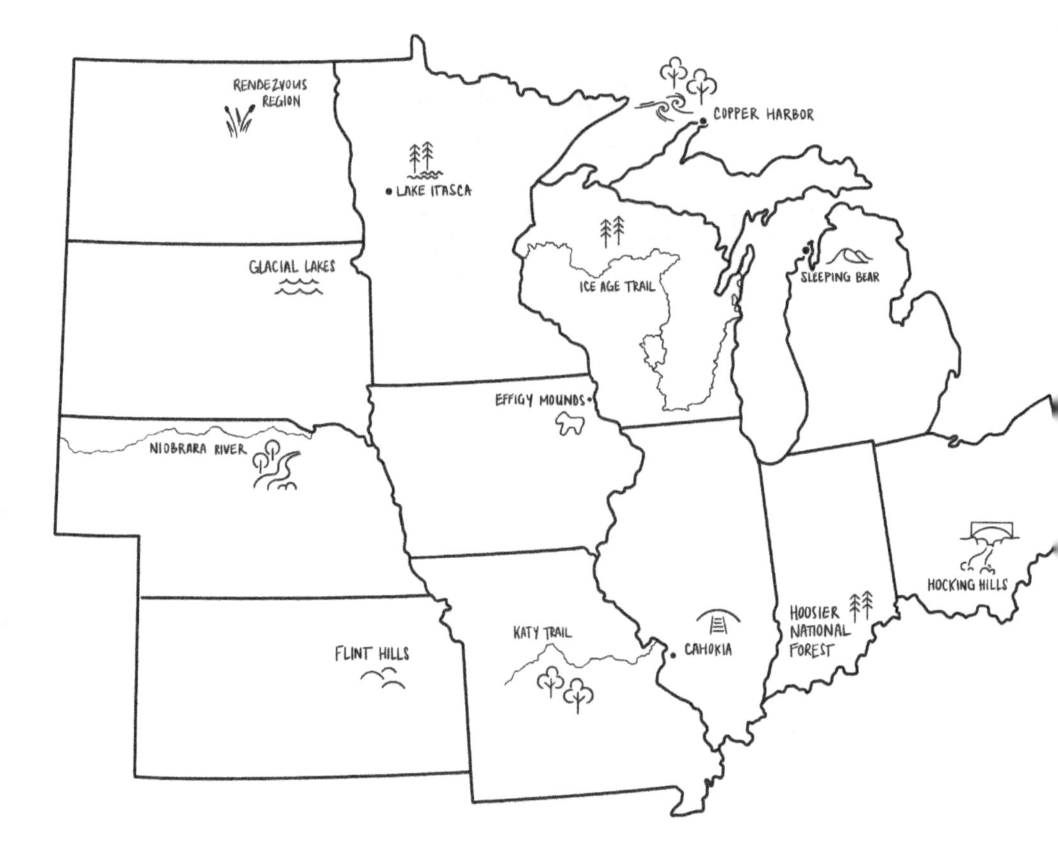
RENDEZVOUS REGION
LAKE ITASCA
COPPER HARBOR
GLACIAL LAKES
ICE AGE TRAIL
SLEEPING BEAR
EFFIGY MOUNDS
NIOBRARA RIVER
HOCKING HILLS
HOOSIER NATIONAL FOREST
KATY TRAIL
CAHOKIA
FLINT HILLS

# Introduction

*Midwest Bedrock*

Like many midwesterners, I spent my youth denying I was midwestern. We all go through this phase. Other places just seem so much more exotic.

But my denial was of a different sort. I knew instinctively that this little pocket where I lived—Dubuque, Iowa, along the Mississippi River—was not the typical Midwest. This I could identify with. This was certainly not the Midwest I thought I knew when I traveled even short distances beyond the Mississippi valley.

I can't put my finger on when it happened. I'm sure it was quite gradual. Perhaps it was when I began dating my wife Dianne more than forty years ago, as we quickly fell into the habit of spending time outdoors together and seeing new places, even if just a short drive from town. When our first son, Paul, was born, we bundled him into a back carrier and hiked amid the autumn gold of birch leaves at the Mines of Spain along the Mississippi. I remember the white trunks etched against the blue sky. After our second son, Brian, came along, we perched them both in child carriers on our bicycles and rode the former railbed on the Heritage Trail, snaking west across Dubuque County. That carrier was on the back of my bike so long that I still dismount so as to not swing my leg across a phantom child seat. When our daughter, Angie—our third and youngest—arrived, Dianne and I were finally outnumbered, but we still took our family chaos hiking and camping. Once, at Pikes Peak State Park about an hour from home, we unloaded and set up camp, and *then* discovered that Angie had brought no shoes, sandals, or flip-flops. Who wears shoes in the summer?

We began to discover not only a broader Driftless Area, but a more varied and interesting Midwest than I had ever thought existed. We scaled the blocky, purple quartzite bluff five hundred feet above Devil's Lake in southwest Wisconsin. We canoed the frisky Wapsipinicon River through eastern Iowa. We camped at Iowa's Backbone State Park where we played in the sandy creek bottom and hiked "the Backbone," a hogback ridge, to its narrowest point where it overlooks meanders of the Maquoketa River on either side of the ledge. Over the years we spent summer weeks in fishing resort cabins in the Wisconsin north woods and Minnesota lake country.

And I came to terms with the Plains, too, as I imagined the weight of mile-high glaciers pressing down, crawling southward with a slow certitude, and scraping the landscape flat. I imagined the glacial till sprouting a few grasses as the glaciers retreated, and those grasses blossoming into the prairie that produced and protected the deep, fertile soils that grow corn and soybeans today.

Our outdoor tastes weren't limited to the Midwest. Dianne and I camped at Rocky Mountain National Park on our wedding trip. As the kids grew older, we also took them to Rocky Mountain, and to Yellowstone, to the Badlands, the Grand Tetons, Yosemite, and the John Muir Woods. When they were young adults, they spent a week with me in Ireland while I was teaching in Dublin. Dianne and I explored Ireland, Germany, England, and France, spending as much of our time as possible on long outdoor hikes and walks in the mountains, around lakes, beside streams, and in hills far away from the famous cities.

The glamorous places—the Rockies, the Tetons, and ocean shorelines—impress the senses with their immensity and grandeur. Nature in the Midwest rarely competes in height, depth, or range.

Midwestern spaces are instead "grace notes." Musically speaking, a grace note is an ornamental sound so tiny its time value is not counted in the rhythm of the bar. As outdoors spaces, the Midwest's grace notes include small county parks with clear, cold trout streams as well as large refuges that are tiny in public awareness.

These midwestern grace notes lie near our homes. We will turn a corner on a winding road in the Driftless to find a rock tower hovering above a meandering stream. On that long interstate drive we will see in the distance the curious crests of Nebraska sandhills, and a detour onto the backroads will reveal to us that lonely yet profound landscape. Appreciating the prairie requires finding the universe in a square yard of ground. These are gift pennies, making us rich only in their moments of sublimity.

And they are bedrock. Literally, in terms of limestone bones of earth that rear up out of the soil in the Driftless. Figuratively, they are the solid physical places in a nation awash in uncertainty.

---

So what is Midwest Bedrock?

I suppose a more basic—or bedrock—question should precede this: What exactly is the Midwest? Geographically, the United States Census Bureau defines the Midwest as the following twelve north-central states: Ohio, Indiana, Michigan, Wisconsin, Illinois, Minnesota, Iowa, Missouri, North Dakota, South Dakota, Nebraska, and Kansas. In common usage, some will rule out Ohio as too eastern, too much like forested, coal-mining, industrial Pennsylvania. Some will drop North and South Dakota as too western. Others consider Missouri too southern, Kansas too south-central. I like the Census Bureau's no-nonsense deliberations. In terms of gathering census data, there are four US regions: the Northeast, the South, the West, and the Midwest. There's something midwestern in that simplicity.

The term seems to have originated out of nineteenth-century US expansion. The Midwest states east of the Mississippi were once simply called the West or the Northwest (hence Northwestern University in Chicago), until the Louisiana Purchase added vast new regions to call the West. Still not quite East and definitely not South, the term Midwest gradually took over.

That is, when it wasn't being called by other names. A different name like "Mid-America" would be less confusing than "Mid" West. Most kindly, some call the Midwest the "Heartland," suggesting not only its physical centrality but also, well, a certain bedrock, plainspoken personality of its people. "Prairie States" is a fairly accurate descriptor of its historical flora, but ignores the vast historical stretches of forest and oak savanna. "Plains States" plays to a topographic stereotype and minimizes the Midwest's varied landscapes while likewise suggesting a rather milquetoast state of mind. What excitement could come from the Plains?

And worst of all is the pejorative "Flyover Country," so-dubbed by eastern and western elites flying between coastal cities. Many on the coasts think they know us: I've tried convincing easterners and westerners that there are considerable hills where I live along the Mississippi River.

The problem lies in the desire to quickly label a region without much thought to its nuances. The Midwest is anything but homogenous. Its landform (or physiographic) regions run from the glacially flattened to the rugged and rocky. On the east it is wet and humid, on the west it is high and dry. The northern midwesterner takes a certain pride in getting buried in snow while the southern midwesterner comes to a halt in the wake of a few inches.

Aware of its diverse landscapes, we call the whole region "Midwest."

So that leaves bedrock.

Most of the Midwest's literal sedimentary bedrock coalesced under ancient seas while the North American mantle drifted over the equator from three hundred to five hundred million years ago. Limestone formed when those seas teemed with shelled creatures whose hulls sank to the ocean floor after they died. Compressed and chemically altered, they exuded lime, which hardened into rock—and created some pretty cool fossils. Sandstone formed in shallower seas near shorelines, as layer on layer of sand packed into stone. Shale formed where muddy rivers emptied into the sea. With the land rising and falling and rising again, a slice down through Midwest bedrock today typically produces alternating layers of limestone, sandstone, and shale.

Northern Midwest bedrock has different origins. The granites and basalts of northern Minnesota, Wisconsin, and Michigan attest to once-grand mountains worn away by time and repeated glaciers.

Try as I might, I can't completely avoid slipping into the metaphorical sense, the somewhat tired cliché of the Midwest being the psychological bedrock of the nation. This book makes no attempt to define a midwestern mindset or culture. Much has been written about "midwestern nice," in terms of both friendliness and a certain passive-aggressiveness in not speaking of difficult matters—except behind someone's back. There is also a perception of the Midwest having a work ethic valued by employers and particularly noticed when midwesterners emigrate to other regions of the country. There may be certain truths to these, but I suspect they are also exaggerated. That said, you may need to excuse me if my vision of "Midwest Bedrock" occasionally suggests a regional personality. It's hard not to love your neighbor here.

This book, then, is a search. Geographically, there is no denying that the Midwest is at the center of the nation. But what is at the center of the Midwest?

---

It is worth stating a few things that this book is *not* about. It is not about the Midwest's politics, but suffice it to say that I have grown increasingly despondent about my home region's political landscape. I once took pride in much of the Midwest's "purple-state" status, but the region has grown more extreme over the last decade. However, such changes don't happen in a vacuum. Declining rural population and increased poverty expresses itself in political ways that—in my opinion—don't always serve our best interests. I also lament the lack of political concern for our polluted midwestern streams and rivers.

Nor is this book about the Midwest's cities, but it might be worthwhile to set straight some misperceptions of the Midwest as excessively rural. Nine midwestern cities are among the fifty most populous in the United States. Below are their national rankings and populations according to the 2020 US census:

3) Chicago, IL (2,670,000)
14) Columbus, OH (914,000)
16) Indianapolis, IN (887,000)
26) Detroit, MI (664,000)
30) Milwaukee, WI (588,000)
39) Kansas City, MO (502,000)
41) Omaha, NE (480,000)
46) Minneapolis, MN (439,000)
50) Wichita, KS (391,000)

Cleveland (OH), St. Paul (MN), Cincinnati (OH), St. Louis (MO), Lincoln (NE), Fort Wayne (IN), Toledo (OH), and Madison (WI) likewise place in the top one hundred.[1]

The twelve Midwest states accounted for 68.8 million of the nation's 331.5 million people in the 2020 census.[2] That means we represent 21 percent of the nation's population in 24 percent of the nation's states, pretty much right on average as a region. So much for the Flyover Country.

Dianne and I enjoy the occasional day in the city, visiting our daughter in Chicago and taking in the museums, or investigating the St. Louis Gateway Arch or the parks of Minneapolis. For the most part, though, cities are what we pass through on our way to more inviting places.

Finally, this book is neither a tourism manual nor a book about the twelve best-known places in the Midwest. Were it that, there would be chapters about Six Flags over St. Louis, the Wisconsin Dells, the Ozarks of Missouri, and Mt. Rushmore of South Dakota. Nothing against those places, but this work seeks to take you places that you may not have heard of or given much thought to. The regions highlighted here may offer readers—even midwesterners themselves—a few notes of surprise.

Who knew that Wisconsin is home to an eleven-hundred-mile national scenic trail, the Ice Age Trail, that roughly follows the end moraine of the last glacier, separating the ice-leveled north from the rugged hills of the unglaciated Driftless Area?

Who knew that Michigan's sixteen-hundred-mile sand-drifted shoreline along Lakes Superior and Michigan is a remnant of the even larger lakes that existed at the height of the glacial melt back? Or that copper mines once ruled Michigan from depths as far as fourteen hundred feet below ground?

Those who have driven across Indiana and Ohio's interstates have witnessed vast agricultural resources on the Plains. But southern Ohio first dimples and then rises and plunges in a region called the Hocking Hills, festooned with box canyon waterfalls. Nearby lies a coal-mining past in the Appalachian foothills in one direction and the birthplace of the Hopewell culture of mound-building Native Americans in the other.

Southern Indiana's Hoosier National Forest fans out across two hundred thousand acres of hilly land reforested by the federal government after early farming erosion and the Great Depression brought about widespread farm foreclosures. This is karst country, too: east of the forest lie great cave networks channeled into the subsurface limestone.

It may come as no surprise that Kansas is grass country, but who knew that eastern Kansas is home to 80 percent of the nation's remaining tallgrass prairie? The prairie there is preserved because the hilly, rocky terrain of the Flint Hills made the land too rugged to plow. So ranchers graze cattle on the nutrient-rich native grasses.

I advise drivers crossing Nebraska on Interstate 80 to detour to the northernmost tier of counties to the Niobrara River, which downcuts through a two-hundred-foot canyon graced with numerous waterfalls. You will pass through the haunting sandhills along the way.

I may live along and be partial to the Mississippi River, but the Missouri River is its chief North American rival. Bicyclists love the Katy Trail that follows along the river valley halfway across the state of Missouri. History is layered here as the trail runs on the bed of the historic Missouri-Kansas-Texas (or MKT, or Katy) Railroad. And layered beneath that is the pathway of Meriwether Lewis and William Clark and their Corps of Discovery as they ascended to the West and returned to St. Louis on the Missouri River.

Natural, Native American, and early pioneer history layers the Mississippi River shoreline along western Illinois from East Dubuque to the Ohio River confluence. The Native American city of Cahokia grew to a population of twenty thousand around AD 1100. Cahokia was the seat of the Mississippian culture where the Missouri and Illinois Rivers joined the Mississippi, offering highways to the north, south, east, and west. Location, location, location.

The North Dakota landscape may look unpeopled, but the northeast corner was once called the Rendezvous Region, where French fur traders and Native Americans came together annually to transport beaver fur and bison hides across the landscape to eastern seaboard and European markets. The flatland that once sat at the floor of Glacial Lake Agassiz is now the vast and frequently flooded plain of the north-flowing Red River. But the Pembina escarpment rises sharply and surprisingly from this plain, leading to a more rugged terrain to the west.

The eastern corridor of South Dakota defies outsiders' perceptions of the state. South Dakota evokes images of eastern plains and the haunting terrain of the western Badlands, Black Hills, and Mount Rushmore. But northeast South Dakota is a lake-filled area called the Glacial Lakes and Prairies Region, and the southeast corner surprises with its orange-pink quartzite arising out of the landscape only to be sliced by downcutting rivers.

The long story of the Mississippi River begins with a small stream emerging from Lake Itasca in northern Minnesota. The infant river wants to flow quietly north, which it does for a while until the Laurentian and St. Lawrence Divides redirect it southward, sending it on a journey that bisects the country and gathers to it the great rivers, so that in the end the Mississippi watershed drains 40 percent of the United States. Nineteenth-century explorers paddled through the North Country making claims and counterclaims as to the source of the great river. But how does one decide which flow leads to the headwaters and which ones are tributaries? The local Ojibwe were bemused by the quest. Didn't all of its flows contribute to the Father of Waters?

The Mississippi River runs beneath the bluffs of my home in Dubuque, Iowa. While relatively little of Iowa's land fits the prevailing mental construct of the state as flat, northeast Iowa in particular shatters the common perception. Steep hills and valleys, rock towers, caves, and freshwater springs pop from every turn along our winding roads in Iowa's corner of the Driftless Area, or what geologists call the Paleozoic Plateau. This was sacred land, a refugium to plants, animals, and humans sheltering from the great ice to the north. In this refugium, Native Americans built two-hundred-foot-long burial and ceremonial mounds, many shaped as bears and birds, on the bluffs overlooking the Mississippi. Here in the northeast corner of the state, Effigy Mounds National Monument protects the largest cluster of effigy, conical, and linear mounds in the country.

Eleven of the twelve Midwest state names derive from Native American words or tribal names, however distilled, mangled, and misunderstood by European and American interpreters. And so we have: Illinois, from Illiniwok, simply meaning "men" but misapplied by Europeans as a name for regional tribes; Iowa comes from the Ioway tribe; Kansas is from the Kansa people, also known as the Kaw; and Missouri is from the tribal name meaning the "canoe-havers." In relation to natural features: Michigan comes from an Algonquin word meaning "big lake"; Minnesota arose from a Dakota word meaning "cloudy water" or "sky-tinted water," depending on the pronunciation; Ohio comes from the Iroquois word meaning "beautiful river"; Wisconsin is from the French interpretation of a name for the state's largest river, the "Ouisconsing"; and Nebraska is close to the indigenous name for the Platte River, *ni braska*. North and South Dakota derive from a term meaning "friends" or "allies." The one Midwest state whose name, ironically, *isn't* derived from an indigenous word—Indiana—is an American-English reference to the Land of the Indians in general.[3]

An introduction is a bit early to debate the ethics of Euro-Americans adopting indigenous place names while simultaneously purging indigenous peoples from those same lands. One side will argue that Whites remembered the Native American past by so naming their states and landmarks, while another side sees the practice as whitewashing history. Yet a third view offers that regardless of the reasons and motivations, if Euro-American founders had not so-named their states, there would have been even more erasure of the Native American past from the landscape.

For the moment, though, let's entertain some different state configurations for the Midwest. I'll name them here for their landscape regions to show that the Midwest is more diverse in its physical geography than most would suspect. My objection to the current twelve state boundaries is that they are political creations often unrelated to the landforms on which they sit. I'll leave it up to you whether or not to give these new states indigenous names.

The Midwest states redrawn according to seven landforms would be:

- Northern Glacial Plain: This state would run across the northern Midwest, including the upper halves of Wisconsin, Minnesota, and north-central Iowa. This is not the only region to have been leveled or otherwise impacted by the most recent (Wisconsinan) glacier, but the glacier's effects are particularly visible on the landscape here in the forms of kettle ponds, drumlins, eskers, and moraines. Much of North and South Dakota would fit here, too, but I'm putting them into different states. You can see the problem already.

- Tallgrass Prairie: This might be the largest of the Midwest states, running north-south through eastern North and South Dakota, east Nebraska, and east Kansas, and then plunging eastward through Iowa, northern Missouri, north-central Illinois, and a bit of northwest Indiana. Here, big bluestem once grew taller than a rider on horseback, and wildflowers bloomed and waved throughout the summer. Very little tallgrass prairie survives: its rich prairie soil has largely been utilized for farmland.
- Shortgrass Prairie: This state would parallel the western flank of the Tallgrass Prairie state from north to south, but its drier conditions would be noticeable in the smaller height and different variety of grasses. These lands today are either irrigated for agricultural use or grazed as cattle ranches. The Shortgrass Prairie state, while relatively level to the local eye, would nonetheless commence an upward slope to the Rocky Mountains. The western halves of North and South Dakota, Nebraska, and Kansas would make up the Shortgrass Prairie state.
- The Great Lakes Perimeter: This state would outline Lakes Superior, Michigan, Huron, and Erie, a narrow and sinewy region snaking the lakes' shoreline through Minnesota, Michigan, Illinois, Indiana, and Ohio. This thin band along the lakes has a landscape all its own, having been the former bed and shoreline of the postglacial lakes.
- Eastern Woodlands: The Eastern Woodlands state comprised of north and central Indiana and Ohio would blend at its edges with the Northeast state of Pennsylvania. Perhaps the definition of Midwest would again be under contention.
- Southern Hills: This east-west state would run across the southern flanks of Ohio, Indiana, Illinois, and Missouri. This is a rugged, hilly, and forested region that lies beyond the farthest reach of the glaciers.
- The Driftless: The Driftless region encompasses twenty thousand square miles covering the corners of four midwestern states: northwest Illinois, northeast Iowa, southeast Minnesota, and southwest and central Wisconsin. Its ancient hilly and rocky landscapes were neither bulldozed nor plowed under by the glaciers. Its soils contain little or no glacial drift from the Wisconsinan period glacier—glacial drift meaning anything dragged along with, dropped by, or blown in from the massive ice, hence "Driftless." This is my state. Viroqua, Wisconsin, already claims to be its capital.

I'll be the first to admit that these boundaries overlap, and that there could be other and different configurations. Perhaps it would have to be settled by political means nevertheless.

But imagine identifying with one's landscape instead of lines drawn arbitrarily across a map.

Imagine identifying as midwestern.

# 1

# Wisconsin

*The Ice Age Trail: Landscape Tells a Story*

## St. Croix Dalles at Interstate Park (Mile 1)

They might have been gigantic witches' brew pots painstakingly ground into the bedrock. The smaller ones could have been mixing bowls and the tiny ones inverted cups for sipping rainwater with a straw.

But geologists call them potholes, circular indentations in the bedrock formed when roaring glacial meltwater on the St. Croix River drained the long-gone Glacial Lake Duluth. The meltwaters burst through a sandstone and basalt lip in the landscape, creating Taylor Falls and a hundred-foot-deep gorge called the St. Croix Dalles in its roaring wake. Potholes formed when rounded stones—as big as bowling balls and as small as marbles—spun dizzily in the eddying currents and downcut perfectly circular holes into the bedrock. Some of the larger potholes were as deep as six feet and as wide as twelve.[1]

The Wisconsin Ice Age Trail was not yet on my mind years ago when I first pulled into Interstate Park on the Wisconsin-Minnesota border to wander amid this curious landscape. Bisected by the St. Croix River, the park resides on both states' rocky shoreline boundary. Winter snowmelt had set the falls roiling and the St. Croix spraying and tossing from one side of the rock-walled gorge to the other. It was an annual, miniaturized replay of the end of the glacial period twelve thousand years ago.

On a bluff overlooking the St. Croix River, as it settles out downstream from the falls, sits another boulder, this one embossed with a plaque announcing the western terminus of the Wisconsin Ice Age Trail. For those willing and able to devote their time and legs, the Ice Age Trail here begins a 1,150-mile trek across Wisconsin, following the approximate edge of the last glacier's advance. Stretching roughly from northwest to southeast and north again across the state, the trail separates Wisconsin's glaciated regions to the north from the unglaciated "Driftless" region to the south. A transitional washout zone from the edge of the melting ice occasionally buffers the two landscapes.

Landscape always tells a story. The grayish-black basalt here at Interstate Park—so different from the Midwest's prevailing limestones, sandstones, and shales—formed from lava flows erupting along fault lines a billion years ago. Half a billion years later, the basalt lay beneath a shallow sea as the North American continental plate slid slowly across the equator. The sea compacted a thick sandstone across the basalt, and eventually the region drifted north and uplifted. Repeated ice ages buried the land under glaciers beginning two million years ago, followed by intermittent thaws. Just twelve thousand years ago, a blip in geologic time, the last great glacier was melting in full force.

I was hiking on the Pothole Trail, two hundred feet downstream from where the meltwater torrents ripped through, creating the Falls, Dalles, and potholes, and scattering ten-foot-high angular blocks of basalt around the gorge. The Pothole Trail is perched high above the St. Croix, indicating that the engorged river ran one hundred feet deeper during the deluge than it does today.[2]

It was a long time ago. I'm referring of course to the lava flows, ancient seas, and glacial melt, but also to my visit to Interstate Park. I was traveling alone, researching another writing project. My wife and older son back home had just picked out a used car for his teenaged needs. They were consulting with me on my first cellphone as I picked along the trail.

I might as well have been wandering amid the dinosaurs, it was so long ago. My son is in his thirties now, and my flip-phone long ago disappeared into another watery tumult: our family's washing machine.

But the stories are still here in the landscape. And the great ice still seems only a grinding crunch away.

## The Wisconsinan Ice Age

Do you hear it coming? The snap of a tree trunk in its path, the cracking of bedrock underneath, the groan of ice crawling in the valley just beyond the bend? You would have to listen and watch for centuries and let yourself be numbed to the bone, of course, but the wall of ice is on the move, ever so slowly.[3]

What set off the great beasts from the North? The earth's tilt cycles every forty-one thousand years between twenty-two and twenty-five degrees off vertical, and when the tilt is greatest, the North Pole will be farther away from the sun in winter. Winter deepens. The earth wobbles like a top through space, and every twenty-two thousand years completes a cycle that likewise puts the North Pole farther away from the sun. Winter deepens. The earth's orbit around the sun cycles between elliptical and nearly circular paths, and when the extreme elliptical path puts us farther away from the sun for long stretches of the year, winter deepens.[4]

When all three cycles align and conspire, the ice may awaken in the North.

Why not in the South? A perhaps not-so-obvious requirement for continental glaciers is the presence of continents. The Southern Hemisphere hasn't experienced widespread glaciers except, of course, on Antarctica itself because there are no other large land masses near enough to the South Pole for glaciers to crawl across.

The Wisconsinan period of glaciation had merely been the latest, beginning one hundred thousand years ago. Twelve thousand years ago, it was in full retreat.

Earlier glacial advances had also covered North America. The Illinoian glaciers conquered the upper Midwest 300,000–130,000 years ago, outlining, as the name suggests, the present-day shape of Illinois. Pre-Illinoian glaciers—sometimes divided into Nebraskan and Kansan—blanketed the northern states even earlier, up to 2.5 million years ago. All of the midwestern states were touched by at least one of the glacial periods.

But only the most recent glacial period is still prominently written on the landscape. Its impact was unmistakable. The Wisconsinan glacier was six hundred feet thick near its edge, but was more than two miles deep farther back over present-day Hudson Bay. Evidence of its movement and melting is still apparent.

The Wisconsinan glacial period also impacted human habitation in North America. So much of the earth's water was tied up in the ice that sea levels fell four hundred feet, opening up a land path from Asia to Alaska (a swath that remained fortuitously unglaciated) that humans followed in their incessant wanderlust. Many lingered at the glacier's edge, hunting mastodons and wooly mammoths until these megafauna suffered perhaps the first human-induced extinctions. Others dispersed throughout North and South America, the ancestors of Native Americans, the first immigrants to the New World.

---

The ice in my driveway that winter seemed solid as a stone at the curb where I'd given up trying to chop it down to the concrete. The ice on my roof seemed immobile, till during a slight thaw it let loose and pulled the gutter down with it.

How, then, on a flat plain, does ice begin to crawl?

When snow survives the summer melt, as it did when the world grew colder and winter grew longer, it compacts into ice beneath the next year's snow, which in turn compacts during the following summer's partial melt. At a hundred-foot depth, after centuries of accumulation, the ice begins to flow like a heated plastic, at first molding itself to the contours of the land, but then—as it gains weight, depth, and force—gouging, scraping, and bulldozing the land as it moves. In one place, the ice might crawl along at a few meters per year, somewhere else one or two kilometers. At its final reach, where summer melt offsets the push of new ice, the glacier's edge doesn't move at all, but the ice behind still presses forth. Stalemate.

The Wisconsinan glacial period saw numerous retreats and advances. Twenty thousand years ago, it began its final assault. The earth's annual average temperature dropped by six degrees Celsius. At the center of the continental glacier, the average temperatures plunged by twenty degrees.

The ice pressed southward through Canada and the northern United States, as well as northern Europe and Asia, often in adjacent lobes that competed for the ground they covered. In Wisconsin, the Superior and Chippewa lobes covered the northwest corner of the state. The Wisconsin Valley and Langlade lobes fronted the north-central regions. The Green Bay and Lake Michigan lobes swallowed the east and southeast.

---

Can you calculate a tipping point? When did the summer grow warm enough and the winter less intense so that the ice groaned to a halt and began its slow retreat? About twenty thousand years ago, the glacier had reached its full extent and started its retreat.

But was it on a Tuesday in April?

Try this. I am in my early sixties, and that time has gone by in the blink of an eye. Count by fifty-year intervals, one per second, a generation and a half each, as your watch ticks away, 50-100-150-200, and so on. In one minute you will cover three thousand years. In another three minutes, you will be back twelve thousand years ago with the area's first humans at the glacier's retreating edge.

Chip off a block and put it in your drink.

## Chippewa Moraine National Scientific Reserve (Mile 169)

At the time of that St. Croix River trip, I was writing a book about the Driftless Area where I live, a hilly region with tall cliffs and steep, wooded valleys along the Upper Mississippi River valley and inland that the bulldozing glaciers had bypassed. But to know the Driftless, I also needed to know ice. So I took off in early April from my home in Dubuque, Iowa, and drove 230 miles north to the Chippewa Moraine National Scientific Reserve in northwest Wisconsin before proceeding on to the St. Croix.

There had been a hint of spring the day I left, winter having been stubborn and persistent till then. But as I neared northwest Wisconsin, the weather took another turn, first with drizzle, then flurries, and by the time I awoke the next morning in a hotel in the small town of Chetek, two inches of wet, sloppy snow had fallen, adding to the accumulation that still had not melted in the ditches and woods. It seemed appropriate that a spring snow squall had accompanied my venture to see glacial deposits.

The visitors' center at Chippewa Moraine sits atop a small hill overlooking a prairie on one side and a deep woods behind. My understanding of glaciers was immediately turned upside down.

Literally. For the hill I stood on, I learned, had been a low spot in the glacier—a glacial lake, to be precise. The surrounding lowlands had been high spots in the glacier. How so?

A low spot on the glacier accumulated debris—boulders, gravel, sand, and soil. At Chippewa Moraine, the depression became an ice-walled lake. The lake bottom accumulated sediment. Then, as the glacier melted and retreated, the sediment that had been trapped in the low spots and ice-walled lakes deposited out as hills (or hummocks) on an otherwise flattened landscape.

Everything within sight offered a glacial story, including the forested ridge just beyond the prairie. Chippewa Moraine had sat at the glacier's edge where the bottom ice was stagnant, as if glued to the ground beneath it. But the center of the glacier kept pushing. Even though the edge was no longer advancing, movement from within the glacier kept thrusting new ice up over the old in piggy-back fashion, creating a jumbled surface. Boulders, rubble, and debris that had been picked up in the glacier's long scrape across Canada and Wisconsin were carried to the edge, and when the glacier melted back, the left-behind debris formed what is called a moraine, now visible as a line of forested hills.

Down in the icy woods behind the visitors' center were kettle lakes formed by huge blocks of ice that had broken off from the retreating glacier. The weight of the calved blocks had pressed down into the newly exposed and still-soft soil, creating small lakes and ponds after the final melt.

Later that day I found the Ice Age Trail segment that leads to and from the Chippewa Moraine Visitors' Center. I hiked a half mile or so through the mix of snow, ice, and standing meltwater, then decided to save the other 1,149 miles for another time.

## Wisconsin Landforms

You can read the landscape once you know what to look for—or at least make good guesses. My drive to Chetek at the northern edge of the Driftless had been through a familiar landscape. But between Chetek and St. Croix Falls, the landscape noticeably changed. Steep, sharp hills gave way to a gentler topography dotted with lakes. The roads were straighter here than in the Driftless, and where they strayed from the vector line it was to dodge a pond, not to follow the contours of a hill.

I had just traversed two of Wisconsin's four landform regions. Each of the four regions tells a story of ice, even those that weren't covered by it. The glaciated Northern Highlands that stretch across the northern third of the state—including my drive between Chetek and St. Croix Falls—are underlaid by igneous and metamorphic bedrock, the remains of mountains on an ancient landmass known to geologists as the Marshfield continent.[5] Worn away by time and repeated glaciers, the remnant igneous bedrock of the Northern Highlands is covered by a sandy, gravelly soil dropped by the glaciers or washed over the landscape as they retreated. The region contains over forty-two hundred lakes.[6]

The Eastern Uplands along Lake Michigan were also glaciated, but to different effect. This region bears the battle scars of competition between the Green Bay and Lake Michigan lobes of the glacier. The battlefront left a line of hilly moraines at their edges and between the lobes.[7] We hiked many of these on our Ice Age Trail ventures.

The Western Uplands, or the Driftless Area, are my southwest/central Wisconsin playground for bicycling, kayaking, and hiking. The hills make this region well drained: there are few natural lakes or wetlands, but instead fast-moving streams and rivers quickly deliver rain, snowmelt, and spring water downstream and eventually to the Mississippi. Although never touched by the glaciers, meltwater rushes etched the landscape even deeper and chiseled steeper slopes down to the major rivers once the meltwaters subsided.

Sandwiched in the center where these three regions meet is the Central Plains, including the bed of Glacial Lake Wisconsin that formed when the Green Bay lobe blocked the eastward flow of the Wisconsin River. This ice-walled glacial period lake was 160 feet deep and eight times the size of Lake Winnebago (Wisconsin's largest modern-era lake). When the ice dam burst, the lake drained to the southwest in a matter of days or weeks and carved out the steep-walled Wisconsin Dells, flooded today's lower Wisconsin River valley with fine sand from the lake bed, and even barreled into the Mississippi, deepening its valleys past my home in Dubuque.[8]

## The Ice Age Trail

It has never been my intent or desire to hike the entire 1,150 miles of the Ice Age Trail. My approach has been to sample it here and there for hikes ranging from an hour to a day, adding segments to my tally like notches on a stick.

The Ice Age Trail is one of eleven national scenic trails scattered across the United States and nominally administered by the National Park Service or other federal agencies. The National Scenic Trails System includes such well-known giants as the Appalachian Trail and the Pacific Northwest Trail.

Established in 1958, Wisconsin's Ice Age Trail is managed by the nonprofit Ice Age Trail Alliance (IATA), and is maintained primarily by volunteers along its corridor. The IATA, as well as nature organizations, private landowners, and state and county parks and forests, owns sections of the trail. Almost eight hundred miles are completed and open to the public, with the remaining miles temporarily delegated to routes along connecting roads.

Landscapes along the trail are infinitely varied, even for someone like me who hasn't hiked every single mile. My wife Dianne and I have been surrounded by tallgrass flatland prairies on the trail. We've paused at three-hundred-foot cliff faces. We've sat at lake fronts. We've slogged through marshes, sometimes on boardwalks, and sometimes, in the thick of a wet woods, we've grown tired of sidestepping puddles and tramped straight through them.

Add in the seasons and the landscapes quadruple, at least. We've sweated in summer sun and marveled at fall leaves and spring wildflowers. We've snowshoed on the trail. (I half-seriously offered to Dianne that, postretirement, we should snowshoe the length of the trail. Not gonna happen.)

On an early winter day, I stopped in and spoke with Mike Wollmer, then executive director of IATA, at the organization's headquarters in Cross Plains, Wisconsin. One difference Wollmer sees between the Ice Age Trail and other national scenic trails is that Wisconsin's never strays far from local communities. "We value the communities the trail passes through," Wollmer says, adding that "meeting landowners and people in trail communities is part of the experience of hiking the trail."[9]

Indeed, thirteen "trail communities"—towns and cities along the path, including Cross Plains—act as unofficial hosts. The trail has been "woven into the fabric of these towns," says Wollmer, citing a factor that contributes to the vast volunteer network that assists the handful of full-time IATA staff.

Over twenty-six hundred volunteers in local chapters contribute eighty-two thousand hours per year maintaining the trail by mowing, clearing fallen trees, restoring habitat, and more. "Volunteers are our eyes and ears, telling us about weather damage and potential properties for sale," adds Wollmer. They advocate for the trail in their communities, increasing its presence on maps and promotional plans.

Although more than two hundred hikers are known to have completed the entire route, including three who ran the trail in twenty-one days or less, most walkers are not interested in records. "You've got to hike your own hike," says Wollmer. Most people walk short segments of the trail rather than going straight through. It can even be good for a quick mental release: "You can yell at a tree and just keep walking!"

Or, like Wollmer, you can marry your wife there.

The memories are all there, still in the landscape.

## Summit Moraine (Mile 395)

Over the years I'd been notching quite a few visits to the Ice Age Trail, impressive only to a nondistance hiker. But the first summer of COVID-19 seemed like a good time to make a concerted attack on the trail. What better way is there to social distance than to commune with mosquitoes and trace the farthest reach of the glaciers on foot amid woods, bogs, and lakes? So, with the *Ice Age Trail Guidebook* in hand, my wife and I drove to northern Wisconsin and picked out a tent site at Veterans Park near a town called Antigo.

A rain squall interrupted our campfire shortly after supper but dissipated quickly. The gravel beneath our sleeping bag pads poked me awake intermittently through the night, and I could hear the earlier rain still dripping and settling in among the leaves around the campsite. I heard coyotes yipping in what I hoped was the distance.

In the morning, at five thirty, with Dianne still asleep, I wandered down to Jack's Lake at the campground to catch the sunrise. The lake was cloaked in a gauzy fog, thicker to the north. To the south, the thinning mists were roiling off the shore like ghost thieves with pilfered loot hustling away at the crack of dawn.

After a scrambled eggs breakfast on the cookstove and some precious coffee, we headed onto the first leg of our journey, the Summit Moraine segment. It must have been raining all week long. The trails pooled and sloshed. Our new waterproof boots quickly gave up.

Glacial erratics—rounded boulders carried by the glaciers over hundreds of miles and tumbled and grinded as smooth as agates run through a polisher—lay about the forest like decorative afterthoughts. A smallish stone sat here along the trail. A boulder the size of a stove lay tucked back within the trees.

We hiked a seven-mile segment through an arboretum of native pines, firs, and hardwoods. We stepped gingerly across a bobbing bog boardwalk that flanks Game Lake, named by card-playing loggers from earlier times who'd meet in cabins here for a night of poker. We planted ourselves on a bench to watch the still waters of "Unnamed Lake."

Unnamed Lake. At first I laughed at the sign, then pondered it. Every culture names its surrounding landscape. Mountains, rivers, and lakes all bear names. It's how we tell the story of our living on this earth. In *Spaces for the Sacred,* Philip Sheldrake writes that space becomes place when it is associated with memory and story: "Place is space which has historical meanings, where some things have happened which are now remembered and which provide continuity and identity across generations. Place is space in which important words have been spoken."[10]

You know the adage: If a tree falls in a forest and no one is there to hear it, does it make a sound? Try this: If a lake is so remote that no one has named it, is it really a place?

No, the place does have a story. Just not mine. Till now.

## Lumbercamp Segment (Mile 400)

The crawling ice scrapes across the soil and scratches the bedrock. It wrenches a block of granite from its base in Canada and tumbles it hundreds of miles along in the advancing ice, then drops it in some random place when the ice begins to melt.

Dust and gravel and minute debris sort out along the way. Here is a gravelly hill scraped and shaped into place by the advancing ice and left behind in its retreat. There, a flash washout zone.

Sometimes glaciers tell the story in words the size of boulders and the length of a ridge. We connect the words into sentences with our legs. But the spoken voice of its advance and retreat is long gone. Maybe there is some echo of it over the next hill.

We hiked the Lumbercamp Segment east to west, descending first among dried kettle ponds, and then picking through an emerald jaggle of rounded, moss-covered glacial erratics. Through the mosquitoes, ferns, and muddy puddles and through a tangle of fallen timber, it felt more like a trek through the tropical jungle than the edge of a glacier, but that was just a misreading. As we climbed from the boulder-strewn lowland to a pebbly ridge, it was clear: this is what the glacier left behind.

This is what the glacier wrote, like a book whose author has departed.

## Dells of the Eau Claire River (Mile 456)

Like the St. Croix Dalles, the Dells of the Eau Claire River near Wausau in central Wisconsin are angular volcanic rocks cut through, tossed, and tumbled by glacial meltwaters from the retreating Green Bay and Langlade lobes. The Eau Claire River, not lava, was spilling through the canyon on our visit, flush with last week's rain. The path of the Ice Age Trail alternated between canyon overlooks and stream level cutouts.

We had first visited the Dells of the Eau Claire when our son Brian was working his first professional job in nearby Wausau. His big dog, Achilles, was with us. Achilles was a 150-pound bull mastiff, gentle in his spindled, massive coat (and named so that Brian could say, "Achilles, heel!"). Achilles was struggling in the heat of that summer's day as we explored the glacially scratched rocks, and finally Brian hung back with the dog as Dianne; our daughter, Angie; and I continued hiking.

It was the first indication of a lung tumor that would take Achilles's life a few months later. Brian was devastated. The big dog had been his rock so many hours from home, through a tumultuous romance and a frustrating job. Achilles's death signaled that it was time for Brian to move on. Everywhere he's lived since, Achilles's ashes have rested on his mantle.

When Dianne and I revisited the Dells this time, years later, the memory of Achilles hung about the site, proving, to myself at least, that the spirit, if strong enough, can inhabit many places at once. Proving that memory is etched in the landscape like so many scratches in the rock.

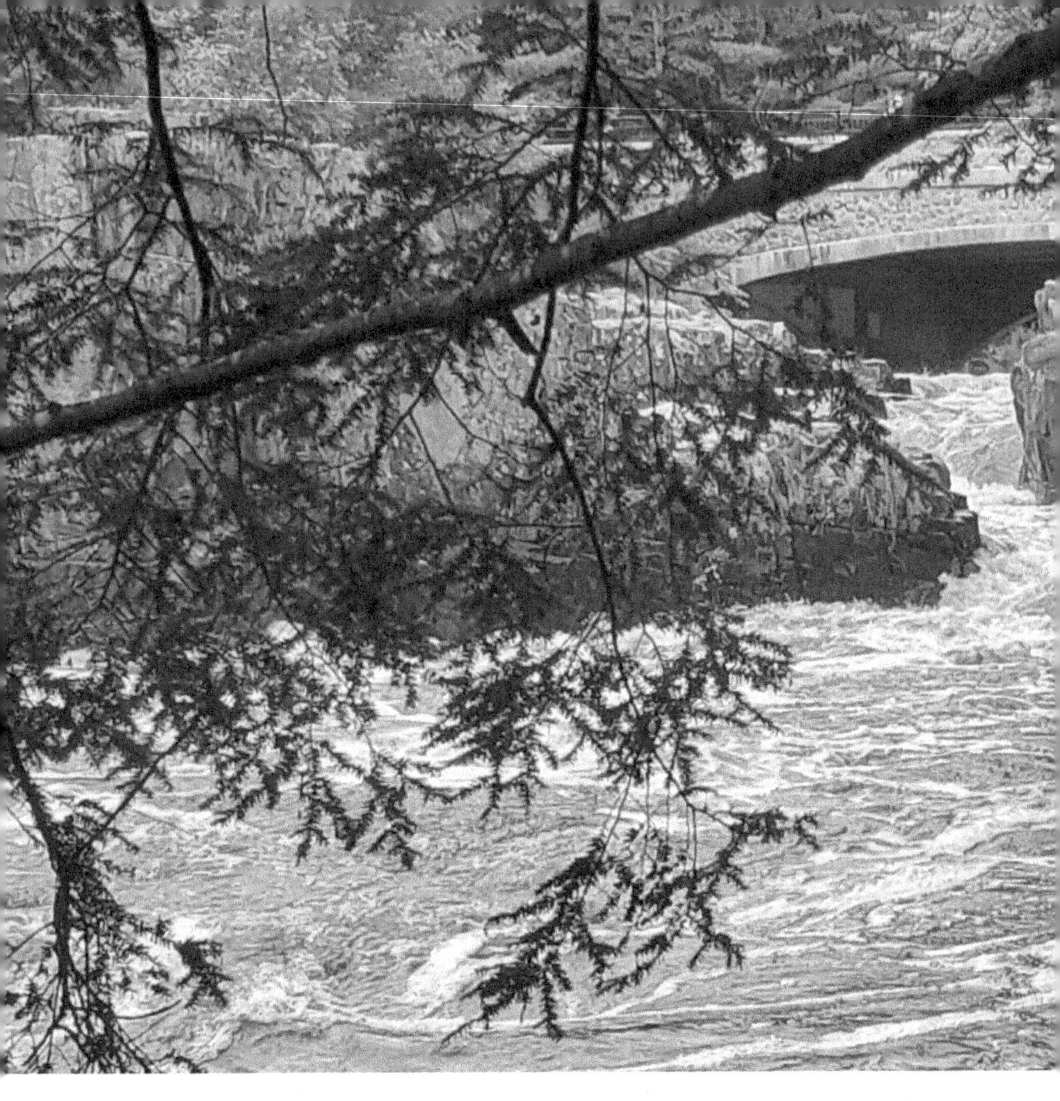

Wisconsin's Ice Age Trail is an eleven-hundred-mile National Scenic Trail that follows the edge of the end moraine of the last glacier from northwest Wisconsin through the center and south and terminates at Lake Michigan in the northeast. Here the trail passes alongside the Eau Claire River Dells in central Wisconsin.

## Devil's Lake (Mile 671)

In south-central Wisconsin two competing snouts of the Green Bay lobe pivoted around an immovable butte and then squared off and faced each other across a narrow valley between eastern and western bluffs before sheepishly retreating. Their high-piled end-moraines sealed off the valley to the north and south and trapped Devil's Lake, with nowhere to run. The

lake is deep and clear, plunging to depths of fifty feet in the steep, narrow valley.

The Ice Age Trail does a near-loop through Devil's Lake State Park. It enters from the east and crosses the northern moraine, then swoops down along the lake's unglaciated west bank, crosses the southern moraine, and heads back again to the east, where it climbs five hundred feet up a spectacular talus pile of purplish, stove-sized quartzite boulders, rounded potholes, and contorted stone formations.

Half glaciated and half in the Driftless Area, the geological story of Devil's Lake begins beneath the sands of an ancient, shallow tropical sea. Ocean-bottom sands compacted into sandstone that metamorphosed into quartzite when continental collisions heated the rock and thrust it upward from a retreating sea. The purplish quartzite bluffs splintered into angular boulders during the intense freeze-thaw cycles of the much later glacial periods.[11]

Oddly contorted quartzite formations perch on the bluff face, five hundred feet above the valley. Most famous among these is the Devil's Doorway, a rock tower with a twenty-five-foot-tall rectangular hole in the center through which visitors of the more daring sort like to scramble.

The base of the scree pile at the bottom of the bluff offers cool relief after the strenuous climb and descent. The gaps between the fallen rocks harbor cool air from the previous winter that slowly leaks out during the summer, creating a cool microclimate at the edges where plants of a more northerly nature thrive.

Native American burial mounds dot the park as well. The earliest peoples arrived twelve thousand years ago as the glaciers were retreating, and the oldest of the circular, linear, and effigy mounds date back two thousand years. One of the most impressive mounds is a birdman effigy with 150-foot wings and human legs and head.

This is where the glacier ground to a halt. Having never reached the west bank and unable to conquer the purplish quartzite east bluff, it gave up its spirit on the north and south moraines.

## Parfrey's Glen (Merrimac Segment, Mile 673)

Just east of Devil's Lake, the Ice Age Trail settles into a more subdued landscape, swatting out across a thick, flat prairie lush with big bluestem, compass plants, and coneflowers, and teeming with grassland birds and insects.

On an early winter hike, Dianne and I took a side trail—as Ice Age Trail hikers are wont to do—into Parfrey's Glen State Natural Area. Euro-Americans settled this glen, or valley, in the mid-1800s, finding that the fast-moving Parfrey's Glen Creek offered power for an assortment of saw and grist mills.

Hiking upstream through Parfrey's Glen is like pacing out a song that starts softly with violins, builds steadily, and crescendos with crashing cymbals, timpani, and bass drums. The first quarter-mile edges gently enough through the prairie alongside a kittenish stream that tumbles out of the forest and hurries across scatterings of well-worn stones.

But the serene bottom-glen is deceptive. The change begins gradually, with the forested creek edge giving way to small outcrops that the stream burbles past. From there, by twist and turn, the path leads upward into an increasingly steep, splashing, Jurassic-Park-like gorge of purplish quartzite.

In its uppermost reaches, the rock walls of the glen are hundred-foot sheer drop-offs carved into the angular bedrock by glacial meltwater. Potholes dot the ledges.

A waterfall feeds the uppermost reach of the gorge where the creek plunges from a quiet forest into the glen.

Our hike wasn't wall-to-wall people, but it wasn't solitary either. At the lower reaches of the valley we watched a boy and his sister hopscotching across the stream atop a few exposed boulders. Farther up the gorge, young couples crawled across the ledges and leaped across stream crossings. While an early winter wind blustered outside the park, inside the glen it felt like a new vernal season had just begun.

## Ice Age Trail Hiker

Barb Shaffer and her husband Danny have logged 443 miles of the Ice Age Trail. A veteran backpacker and former Sierra Club trip coleader, Barb has hiked with groups in the Gates of the Arctic and Denali National Parks, and the Superior Hiking Trail, among others, and even solo-hiked in Michigan's Porcupine Mountains.

She first learned of the Ice Age Trail in 2012, the year before she married Danny and two years before they moved to Wisconsin. "A long national scenic trail close to our Midwest home? Gotta do it!" she recalls saying.[12]

Barb and Danny's first strategy for segmented hikes was to plant bicycles at the end of their intended route so they could bike back to the RV they'd left at the starting point, thus adding some exercise to their exercise. For longer, weeklong hikes, they'd lighten their carrying load by driving first to a camping spot midway along the route and hanging a bear bag from a tree, where it would await them when they reached it on foot. They've also utilized "Trail Angels," local trail supporters who give Ice Age Trail hikers a ride back to their starting point.

Barb doesn't sell the Ice Age Trail short in comparison with other long treks. "This trail is similar to hiking in Alaska due to the areas of hummocky, wet terrain in northern Wisconsin," she said. "Other parts of the

The Ice Age Trail climbs the sandstone butte at Gibraltar Rock within view of the Wisconsin River in south-central Wisconsin.

trail are much the same as other national trails—rocks, roots, and trees leading to beautiful views. I believe it is just as challenging too," she added, notwithstanding the lesser elevation changes on the Ice Age Trail.

Four hundred-plus miles leaves memories on the landscape as assuredly as glaciers leave behind boulders. Barb recalled hiking downstream from a massive beaver dam and then finding herself at eye level with the upstream pool. She remembered bicycling back to the RV with their chocolate Lab, Bella, heeling alongside near the Chippewa Moraine. She recalled thankfully finding a supply of cut and stacked firewood at a hikers cabin at Jack Lake after a wet, muddy day on the trail.

## Gibraltar Rock (Mile 677 Ferry Crossing, Mile 682 Overlook)

A half-dozen raptors rode the thermals just beyond the two-hundred-foot sheer cliff face at Gibraltar Rock State Natural Area near Merrimac, Wisconsin. This, no doubt, was the best way to take in the views of the sandstone bluff and the valley farms beneath, as well as a slice of the Wisconsin River in the distance.

The Gibraltar Rock segment of the Ice Age Trail pads upward along a slightly twisting path softened by needles dropped from towering white pines. Sand ground up and dropped by the glacier that once topped the two-hundred-foot bluff also cushions the soil.

Gibraltar Rock itself is a sandstone butte, formed in shallow Ordovician seas 450 million years ago and then uplifted. More tightly compacted in comparison to neighboring sandstones, Gibraltar Rock held together when water and ice eroded the surrounding bedrocks, resulting in the two-hundred-foot cliff face.

Ten thousand years later, Dianne and I enjoyed the results, eating our packed lunch at the cliff overlook while watching raptors glide in the updrafts. We sat on a natural stone bench a safe and comfortable ten feet from the drop-off, but wondered at the scattered, gnarled cedars that overhung the cliff face like reckless gawkers.

Could you have seen it coming from here, that grinding wall of ice advancing from the north? Slow and slogging as it was, would it appear closer this month than last? Would you hear the snap of a tree plowed under? When would you decide to move your family to new ground?

This day, of course, was warm and sticky. Looming on the horizon were the Baraboo Hills—a thirty-mile line of purplish quartzite rising up to seven hundred feet above the surrounding plain—that had been surrounded but not subdued by the glacial ice.[13]

After our hike, we took a short drive to watch cars, trucks, and motorcycles disembark from and load onto the cable-drawn Merrimac Ferry that crosses the Wisconsin River and returns every fifteen minutes from mid-April through November. Listed in the National Register of Historic Places, the ferry dates back to 1851. It is the last remaining ferry among five hundred that plied across the Wisconsin River at various times and locations. The state of Wisconsin has operated the ferry since 1933. The price is right: vehicles, bicycles, and foot passengers cross the river for free. The ferry is the official pathway across the Wisconsin River for through-hikers on the Ice Age Trail.

Back on the road, we drove the short distance to the south shore of Lake Wisconsin, a seventy-two-hundred-acre reservoir on the Wisconsin River impounded by the Prairie du Sac hydropower dam. The dam, southernmost of twenty-five hydroelectric stations on the Wisconsin River, has been generating electricity since 1914. Today the lake above the dam is a haven for boaters and fishermen, kayakers, and canoeists. Below the dam, the Wisconsin River flows naturally, dam-free and unobstructed, for ninety-three sandy miles from Prairie du Sac to its confluence with the Mississippi.

Lake Wisconsin may be an artificial wide spot on the Wisconsin River, but not far upstream once sat the ice-dammed Glacial Lake Wisconsin. For those who knew the slow, relentless advance of ice, the wall of water after the ice dam burst must have seemed a swifter god, unannounced save for its roar as it turned a bend in the valley, too soon upon you.

## Table Bluff (Mile 714)

An overlook from the trail at Table Bluff, located near Cross Plains in southwest Wisconsin, peers two hundred feet down across a gravelly outwash valley formed during the melting of the glaciers. For we were near the divide where parallel bluffs mark, on one side, the end moraine of rubble and, on the other, the defiant limestone outcrops of the Driftless Area that never submitted to the ice.

My hike at the Table Bluff segment resulted from IATA director Mike Wollmer's advice on how to fill the remainder of an afternoon after I'd interviewed him. The trail here runs 2.5 miles through a mix of open prairies, oak savannas, stream valleys, and steep forest ravines. Big bluestem grasses top out above eye level on the highland prairies. A hardwood forest snuffs out the sunlight where it swallows the trail. The trail lulls a while through the woods before switchbacking in swift descent to the wide glacial outwash valley and ascending again into another thick forest with impressive overlooks.

Fifteen thousand years ago the valley rushed wide and deep with runoff as the nearby glacier began melting back. When the meltwater receded, it left behind a thick layer of pebbly sand. Today's remnant creek gallops along through a thin slice of the valley.

The Table Bluff segment of the Ice Age Trail connects to the town of Cross Plains via low-traveled Scheele Road, one of several such on-road segments. Maybe one day the trail will be entirely off-road, but in the meantime, says Wollmer, hikers say the backroad segments are part of the overall experience where they often meet adjacent landowners.

## Cross Plains, Overlooking the Driftless (Mile 732)

The trail enters Cross Plains along Main Street and passes alongside several local bars, restaurants, and coffee houses before pausing at IATA headquarters.

A short hike through the town of Cross Plains and down a long residential lane leads to an unobtrusive Ice Age Trail sign pointing into the woods and a quick, switchback ascent back into the wild. The path winds through restored prairie at the Andersen Preserve and through an oak savanna, a Driftless Area anomaly that features intermittent stands of ancient burr oaks towering above a prairie floor. Prairie fires once kept midwestern grasslands free of trees, but burr oaks—known for their long, outstretching lateral branches—were more fire-resistant than other species. The resultant oak savannas formed the Driftless Area's transitional zones from the prairie to the deep woods.

Cross Plains took its name from two pioneer roads that intersected here, one a military route linking Fort Crawford on the Mississippi River to Fort Howard in Green Bay, and the other a road running from Madison to towns near the Wisconsin River.[14] But on this segment of the Ice Age Trail, the crossing might have referred to the glacier's halt here in the face of the Driftless hills. From a bluff on the Cross Plains segment the south and east horizons offer up "The Great Dividing Ridge," as conservationist Increase Lapham called the line of hills demarking the edge of the Driftless.[15]

The Driftless—also called the Ocooch Mountains by the indigenous Ho-Chunk—is a landscape of mystery. Its winding roads sink into deep, sharp valleys cut by streams that once carried meltwater from the nearby ice. Limestone rock towers balance in precipitous separation, sheared from the massive, naked bluffs nearby. This is a karst landscape where water has dissolved underground cave pathways through the bedrock. An open stream may disappear into a sinkhole, travel through the cave paths, and emerge as a spring miles away. Algific talus slopes—where porous limestone sits atop impermeable shale—breathe out cold air in summer and suck in the warm. They reverse the process in winter, breathing out warm air and sucking in the cold. This results in microclimates slightly warmer than surrounding lands in winter and cooler in summer, offering habitat to plant and animal species not found elsewhere in the Midwest. I once threw a handful of powdery snow across one such algific talus vent and watched it puff the snow about like confetti. I thought I'd found the hillside where the earth breathes.

What spared the Driftless is a matter of geological speculation, but the causes may have been threefold. The hard granite uplands of northern Wisconsin—the remnant bedrock of ancient mountains—slowed the flow of ice so that it lost momentum and ground to a halt in southwest Wisconsin. Then the lowland scoop of the Lake Superior basin may have chaneled the flow of ice west of the Driftless, and the lowland scoop of the Lake Michigan basin diverted the ice flow to the east.

Older glaciers even swung around south of the Driftless, but never covered it. Through it all, the Driftless remained an island in a sea of ice.

The Driftless Area is my home, although I live across the Mississippi in Iowa. My wife and I kayak in streams and Mississippi backwaters here, and bicyle along winding backroads. Our children, now grown, were teethed in hiking carriers in Driftless parks and refuges. I take my students on winter hikes on bluffs above the river. On solstices, equinoxes, and Celtic feasts, I watch the sun rise over the Mississippi River, cresting above Driftless bluffs. In spring, thin mists hang over the waters. Bald eagles swoop in summer sunrises. Fall is the season of geese barreling out of the fog like airborne bowling pins, and in winter the river rolls unnoticed and silent beneath the ice.

It is the region of mystery.

Southwest and central Wisconsin hold the lion's share of the Driftless, and here today on the Ice Age Trail, at Cross Plains, I stand on the final moraine of the glacier and look out over the outwash plains to the first line of Driftless hills. Beyond that line lies home.

## Devil's Staircase (Mile 816)

Our daughter-in-law Emily's parents invite us to spend a summer weekend at their home in Janesville, Wisconsin, along with Emily and our son, Paul. Emily's father takes us hiking along the Devil's Staircase, a section of the Ice Age Trail that descends sharply from shoreline bluffs down to the Rock River.

I think of Black Hawk and his Sauk followers frantically fleeing up this river valley during the Black Hawk War in 1832. The war—a chase, to be more precise—had begun one hundred miles to the south where the Rock River enters the Mississippi. There Black Hawk and a splinter group of fifteen hundred Sauk defied a fraudulent treaty that had banished them from their home village of Saukenuk on the east side of the Mississippi, today's Rock Island, IL. Recrossing the river to resettle the home village, Black Hawk anticipated support from other tribes, and when none materialized, he attempted a surrender. The surrender, though, was misinterpreted, and a skirmish ensued instead. The war was on. Or the chase. For the next several months of spring and summer 1832, Black Hawk and his band of Sauk hurried up the Rock River with the US Army always a day or so behind, moving so quickly there wasn't much time to hunt and procure food. Sixty miles north of here, with his people starving at Horicon Marsh, Black Hawk recognized the futility of defiance, and routed the Sauk westward to supposed safety on the far shore of the Mississippi. Instead, the Sauk were met by massacre at the river's edge.

But here on the Rock River, in 1832, the pursuit is still on. The boats plying the water today are skimming across another landscape of memory.

Back at the house, Emily and her father walk us along a winding, wooded ridge at the edge of their property. Emily—long before she met and married Paul—had been a student in one of my writing classes, and had written about this very hill and woods. In these woods, she wrote, "I am ageless because my journeys started in preschool and continued throughout my childhood until high school."[16]

The landscape itself is ageless. Seeing the hill slope away on either side, I wonder in my amateur geologist way if they live alongside an esker, a snaking hill formed from sediment dropped from a meltstream at the base of the glacier. Here—*here!*—not some abstract *somewhere,* the great ice left its mark.

## Kettle Moraine State Forest—South Unit (Mile 879)

The slopes of the lightly frosted path dropped steeply away into a stark winter woods. The narrow, curving trail was an esker in the Kettle Moraine State Forest's South Unit in southeast Wisconsin. I remember the landform's name by means of a visual alphabet: its twists and turns remind me of an "S-Curve."

Landscape is a text we can read. Stories accumulate here. I tick off glacial formations like squares on a bingo card as we hike through Kettle Moraine. First, a kame shaped like the bottom half of an hourglass sandpile and formed much the same way from sediment dropped when meltwater plummeted down a vertical chute in the ice. Then I see a teardrop-shaped drumlin, fashioned from rubble dragged and scraped across the bedrock, with an arrowed edge pointing forward in the direction of glacial crawl.

Plus there are two features for which the state forest is named: its glacial moraine and kettle ponds. Such is a reading of Kettle Moraine South Unit, a twenty-two-thousand-acre state forest that stretches in a northeasterly band for thirty miles beyond Whitewater, Wisconsin.

Dianne and I hiked the Kettle Moraine segment of the Ice Age Trail on a sunny but chilly winter day. The winter thus far had been mild, but a stinging wind swiping down from the north offered the imagination a hint of a glacial chill. But only a hint. In truth, the slight dusting of snow served mostly to offer some color contrast to the browned-out woods and brought the sometimes hidden glacial features into relief.

In addition to its glacial features, the park's oak, pine, and aspen forests are punctuated with prairies, springs, and marshes that offer habitat to coyotes, foxes, Cooper's hawks, and sandhill cranes. Kettle Moraine is an Audubon Society Important Bird Area (IBA), home or breeding ground to 137 woodland, grassland, and marshland species.

More of the Ice Age Trail climbs through the Lapham Peak Unit of the state forest, rising from a lowland boardwalk crossing over a marsh and scaling up through the woods to the peak. The trail passes alongside the forty-five-foot wooden watchtower capping the highest point in Waukesha County.

But the land holds in its pages human stories as well and reminds us that our every footprint is inscribed in the landscape and eventually swallowed by it.

A hike through the Scuppernong Springs Trail near the north of the forest's scattered holdings tells how humans have lived in this landscape over time. Archaeological findings of arrowheads and flint flakes point to Native American encampments on high ground overlooking the marsh, a choice location with access to plentiful game and fresh water.

White settlement altered the landscape, and then receded into the background. The one and one-half-mile trail brushes past the fading traces of an 1846 sawmill and the still-standing walls of a 1909 marl plant. Marl is a lime-based, grayish-white clay formed at the bottom of glacial lakes and used as fertilizer and building mortar. For six years, sixty workers dug the clay out of the marsh, processed it, and loaded it onto train cars on rail run specifically to the plant. By 1915 the plant was closed, the railroad abandoned, and the entirety left for the woods and marsh to reclaim.

Nearby, a nineteenth-century cranberry bog and a trout hatchery flourished behind man-made dikes. These were removed in the 1990s to restore natural habitat along the Scuppernong River for native wild brook trout, beaver, otter, muskrat, and mink. The springs themselves still bubble up from the clay-marl bottom.

At nearby Paradise Springs, additional natural springs erupt where the water table is sliced diagonally by the rocky slopes. Thirty thousand gallons of fresh water pour every hour from the rocks at a constant forty-seven degrees, year-round. The resources and scenic valley attracted both entrepreneurs and those seeking an idyllic escape from Milwaukee, about forty miles away. A half-mile leisurely trail—much of it handicap-accessible—winds past the remains of a 1920s horse track and a water bottling plant that once produced the label "Lullaby Baby Drinking Water." The plant's foundations and cement stairway are all that remain.

Beyond these ruins lies the Fieldstone Spring House built in the 1930s to protect the spring waters. The Spring House originally sported a copper-domed roof. Roofless today, all that survives are fieldstone walls and the spring as it emerges from the rocky hillside. The dammed-up pond still harbors trout. The grounds around the Paradise Springs Hotel, once a popular honeymoon resort with ruins located near the pond, had boasted a menagerie of peacocks, monkeys, and pheasants. All are gone now, except for the spring, the pond, and an assortment of ruins and foundations.

Historians speak of palimpsests, ancient and medieval texts having a base—whether parchment or animal hide—so valuable that it was oft-reused. Writers scraped old lettering from the pages to make way for new text, but the old still showed through faintly.

Landscape is a palimpsest. Stories as ancient as the glaciers and first peoples, and as recent as abandoned factories and fisheries, sink back into—but still faintly grace—the land.

## Through-Hiker

Luke Kloberdanz jokes that he was experiencing his "quarter-life crisis" in 2003 when he completed a through-hike of the Ice Age Trail. Still fairly fresh out of college and with a few years of teaching under his belt, he was still asking himself, "What do I want to be when I grow up?" With no suitable answer in mind, Luke said, "I took off walking."

With no cellphone nor even a watch, "I was totally immersed in the daily routine of the walk, following the flow of the natural world," Luke says. He'd wake up with the birds, stretch, do some yoga, eat a light breakfast, drink some water ("important to keep well-oiled!"), plan the length of his hike, and "start walking, it's that simple."

The pace of hiking appealed to Luke and helped him learn to read the landscape. "When you move through the trail at a slower speed, you gain an appreciation for the impact of the glaciers." Looking at erratics strewn about the woods, "You understand how the glacier was a conveyor belt, a boulder train."

He often hiked up to twenty-five miles a day, spacing his trek to arrive at a campground or a town, where he'd check in with the local police department to see where he could camp.

Luke also arranged supply pickups from his wife. "People at that time didn't know much about the Ice Age Trail." Still, he often felt the "goodness of people watching over me." One hot day while he was hiking a road section of the trail, a family car pulled over and someone handed him a cold Coke. People he'd never met before invited him into their home overnight, too. After one particularly brutal twenty-mile hike, he arrived early at a crossroads where his wife would later be dropping off fresh supplies. A man in a red Chevy truck came along, saw him sitting at the side of the road, pulled over, left the truck running with the driver's side door open, and sidled over to talk. "We talked for over an hour about my trip and all the things you're not supposed to talk about: politics, religion. When he left I just cried until my wife showed up. A big ball of stuff I'd been pushing down came out. I was walking on a cloud after that."

The hike took forty-seven days, including five days off.

Luke had some experience in distance hiking before his Ice Age Trail trek. He'd done a through-hike on the Superior Hiking Trail two years prior. He knew how to pack for a pared-down existence. Food, water, inflatable mattress, tent. No cookstove. He carried his trail maps, his journal, and a book. Each night he'd write in the journal, each morning check the maps. He never opened the book.

His backpack weighed thirty-two pounds. Over forty-seven days he lost thirty-five pounds in body weight, just a hair more than the contents of his pack.

Luke went back to teaching after the hike. As the next summer approached, he and another colleague thought about the kids—third through fifth graders—who would be needing summer remedial instruction, and how self-defeating it would be to put them back in desks for the summer when they'd not succeeded there during the school year. So Luke and his collaborator organized a summer program called Saunters in which they and some high school mentors took the kids hiking on different sections of the Ice Age Trail, up to eight miles a day for five days straight, and used the guidebook as their textbook for reading, math, history, and science.

The Saunters program drew the attention of the Ice Age Trail Alliance and drew Luke to the organization as well. By 2013, he'd joined the organization as Outreach and Education Manager. During the 2019–2020 school year alone, IATA was on track to work with ten thousand fourth graders before the pandemic hit.

Since then, Luke transitioned first to Director of Philanthropy and is now Executive Director and CEO of IATA. "I definitely draw upon my own experiences walking the trail when speaking with donors and writing grants."

The Alliance has grown along with Luke. When he did his through-hike, only ten persons had preceded him. Now there are about 250 thousand-milers. And COVID only increased appreciation and support of the outdoors. The number of contributing members grew from four thousand to fifty-five hundred during the first year of social distancing.

Luke doesn't let his day job keep him off the trail, although it helps that the trail passes right alongside IATA's headquarters in Cross Plains. "Hardly a day goes by when I'm not physically on the trail," he says. "I don't know anyone who's come back from a hike and said, 'I've had a bad day.'"

## Kettle Moraine State Forest—North Unit (Mile 979)—A Winter Long Ago

Fishtailing across the highway as Dianne and I plunged into an ice storm seemed an appropriate prelude for our very first visit to the Ice Age Trail many years ago. At least that's what I told myself as I steadied the Honda, tightened my grip on the wheel, and settled in for a white-knuckled drive to Kettle Moraine State Forest's North Unit in southeastern Wisconsin.

The brief and localized ice storm had simply been a vanguard ushering in an arctic blast, and by the time we reached the Kettle Moraine Interpretive Center, temperatures had dipped into the single digits and the wind chill to below zero. With snow pants, multiple upper body layers, facemasks, and scarves, we clipped on our snowshoes and crunched our way onto the Ice Age Trail.

Thirty-one miles of the trail pass through Kettle Moraine State Forest North Unit and, of these, we were content in the frigid temperatures to snowshoe approximately three.

We chose the Butler Lake Trail for our snowshoeing venture. The lake itself is a kettle pond, one of many that give the state forest one half of its name. The trail arches across an esker.

Despite the intense cold, the snowshoeing warmed us while we wandered about the glacial formations. And since the Butler Lake esker winds through a thick forest, rising and disappearing again and sometimes settling back down on the forest floor, we were largely protected against the day's harsh winds.

Summer, no doubt, would have provided a more comfortable and lush experience on the trail, with a multiplicity of plant and animal life encounters. But what better way to have first encountered it than with a little glacial chill? This was the Ice Age Trail, after all.

## Point Beach (Mile 1,054)

It was late in the first COVID summer, and Dianne and I approached the trail's convergence with Lake Michigan. Here, the Lake Michigan glacial lobe had creased up against the Green Bay lobe. As both ice masses retreated, the ground beneath the more-deeply scoured Lake Michigan lobe filled with meltwater. The resultant glacial Lake Nipissing was about twenty feet deeper and significantly wider than today's Lake Michigan, its modern descendant.

A quarter-mile inland, the farthest sandy shores of the former Lake Nipissing are forested and dotted with wetlands. Hiking enthusiasts would be appalled, but for a mile or so I hiked the Ice Age Trail barefoot. On this hot late-summer day I let my feet sink into the accumulation of soft pine needles in a sandy soil, into carpet-like moss patches, and into the dampness at the edges of streams and marshes.

A series of dune ridges gradually arising in progression away from the Lake Michigan shoreline mark successive shorelines of the retreating Lake Nipissing. Stabilized by forest away from the lake and grasslands closer in, the dunes offer habitat to bird species specializing in each. It is also a crossover zone between northern and southern birds, thus making it a premier birding area.

Then, through a break in the vegetation and a slight climb up a final dune, Lake Michigan at last!

We found a section of beach all to ourselves. While Dianne called back home to check in with family, I paced up and down the shore. With each ebb and swell the incoming tide inched higher on the sand, licking across my feet, erasing the stories.

## End Marker at Sturgeon Bay (Mile 1,147)

Dianne and I weren't through-hikers or even aspiring to walk every mile of the Ice Age Trail in segments. Obligations were calling us back home to Dubuque. So I reluctantly settled for paging to the end of the *Ice Age Trail Guidebook* to find a photograph of a massive boulder at Sturgeon Bay adorned with a plaque that reads "Eastern Terminus: Ice Age National Scenic Trail." In the photo, a barefoot toddler leans on the stone for balance.

We've saved for another time the eastern end of the Ice Age Trail. But we've left a lot of stories behind on the landscape in between.

The glaciers have long ago receded. Their etchings are faint and disguised along the farthest reach of ice, where human civilizations have come and gone, repeatedly. We seem hell-bent on melting the last remaining polar ice through climate-changing activities, oblivious to the impact those changes will have on us here at the edge of where the ice once stopped.

Like the child in the photo, we hang on for balance.

# 2

# Michigan

## *A Great Lakes Shoreline: Water, Wind, Earth, and Fire*

Dianne frequently obliges my geology obsession, and so, on our bike ride through the northern Michigan campground, it was she who spotted the poster board perched on the seat of a picnic table with its pasted-on display of regional rocks: agates and fragments of basalt, granite, calcite, and copper. We stopped to look it over.

The campground host noticed us and must have sized us up as the type of people willing to pause along the route, and he struck up a conversation. He kept us paused for half an hour. But it was all good.

We were on the Keweenaw Peninsula, the northernmost tip of Michigan's Upper Peninsula that sits above the state of Wisconsin, largely separated from the rest of Michigan. The Keweenaw Peninsula itself is a fifty-five-mile, sickle-shaped sliver of land that slices into Lake Superior and interrupts the southern shoreline. The northern half of the peninsula is Keweenaw County. This had once been copper country, but the mines had long ago shut down, ghosting many towns. Tourism is king now, although not the waves of visitors who favor the easy-to-reach places. Even so, there is a steady stream of curiosity seekers willing to drive to the end of the earth, or at least this portion of it.

"I stopped in at the gas station last week," the camp host told us, "and the owner said, 'Thank God you're here! I have to run some errands. Could you watch the shop for me for a few minutes?' That's just the way it is around here. There's so few of us, we have to help each other out."

It was so basically elemental, this communal need to look out for one another in a county with a land base that is the largest in the state, a population of twenty-one hundred that's the smallest, and that averages 240 inches of snow each winter.

Oh, there were proper scientific elements and compounds on the poster board rock display—and in area museums—that tell a Michigan story: copper (CU), iron (FE), and quartzite ($SiO_2$). But the elemental basics along Michigan's Great Lakes shorelines were deeper than that, so old they might as well reside in the prescientific elements of water, wind, earth, and fire.

Over the years we'd traveled from our Iowa home to Michigan on several occasions, and almost exclusively along or near its Great Lakes shorelines. The four elements seemed a reasonable way to make sense of what we'd encountered.

## Water

The Black River is in a hurry to reach Lake Superior. Located in the western portion of Michigan's Upper Peninsula in the Ottawa National Forest, it drops two hundred feet in its last two miles along a series of waterfalls. Unlike most midwestern streams that languidly search out the Mississippi River and glide with it to the Gulf of Mexico, the Black River arises on the north slope of the St. Lawrence Divide and dashes quickly to Lake Superior instead.

We accessed five of the waterfalls in succession from trails leading from parking lots along County Road 153. The trails ranged from a short one-eighth mile to one and one-half miles through pine, hemlock, and hardwood forests. There is little understory growth, making the canopied woods feel like a grand ballroom with a ceiling supported by tree-trunk columns. Several of the trails utilize steep, wooden staircases for the final descent into the Black River gorge, where the tannin-tinted flow still downcuts through the bedrock. Gabbro Falls exposes the Upper Peninsula's fiery, igneous past by slashing through black basalt. Great Conglomerate Falls slices through a bedrock formed from small and medium pebbles tumbled and rounded by billion-year-old streams that carried them out to a shallow sea where they were cemented together in a hardscrabble conglomerate rock. At Sandstone Falls, the ancient seashore embedded ripple marks in the bedrock.

The falls themselves range from fifteen to thirty feet in height, sometimes dropping in a single plunge and other times splitting into separate veins or stair-stepping down. Some of the falls are viewable from observation decks, but at others we scrambled out onto the exposed bedrock to watch the river rushing past our feet.

Congress designated the last fourteen miles of the river as a National Wild and Scenic River in 1992. The falls, hiking paths, and roadway constitute a National Scenic Byway.

The final stop is at the Black River Harbor Recreation Area, where the river finally relaxes and eases alongside a line of anchored excursion boats, under a suspension footbridge, past a yellow field of wildflowers, and out into Lake Superior. From there it begins the next phase of its journey leading to the other Great Lakes, through the St. Lawrence Seaway, and out into the wild North Atlantic.

Elemental, elemental. The Michigan story begins in water.

The story of the Great Lakes begins with a series of small, preglacial river valleys that were exploited, gouged, and deepened by successions of advancing ice. The slow-moving glaciers—more than a half-mile thick—scooped down the river valleys, scouring out the Great Lakes basin and later filling them with meltwater when the ice retreated. Lake Superior at its deepest is 1,333 feet; Lake Michigan, 923 feet; Lake Huron, 750; Lake Erie, 210; and Lake Ontario, 802.

But the path from glacier to Great Lakes was neither simple nor direct. As the glaciers began melting twenty thousand years ago, postglacial lakes formed at their retreating edges: Glacial Lake Duluth occupied the western half of today's Lake Superior, and Glacial Lake Algonquin encompassed larger versions of today's Lakes Michigan and Huron. As the glaciers retreated farther and sent forth more ice melt, the three lakes joined as one.

The pathway to today's Great Lakes took a few more bobs and rebounds. The glacial mass had actually pressed down the northern half of the floating continental plate over which it passed, much like one can push one end of a floating log down into water. By eight thousand years ago, the glaciers had largely retreated, and the land—still pressed down and not yet rebounded—drained quickly to the north, leading to lake levels far lower than today's. Commercial shipwreck diver Peter Lindquist found evidence of this low-water period in a stand of ancient tree trunks located 180 feet beneath today's Lake Superior surface. According to Lindquist, "Upright on the lake floor, the trees were one to two feet in diameter and perhaps fifteen to twenty feet high, with branches intact. Sitting on the bottom, it was like looking up through maples."[1]

As the northlands rebounded after the glacial melt back, the northern flow through the lake outlets slowed, and lake levels rose to about fifty feet higher than today. Called the Nipissing lakes, their ancient shorelines appear as high ridges in various locations around the Great Lakes.

The Great Lakes, as we know them at their current size, are only about three thousand years old.

---

We had, as I've said, visited Michigan on several previous trips. In 2011, we sampled Lake Michigan's dunes at Grand Mere and Saugatuck Dunes State Parks. In 2018, we drove north to the Porcupine Mountains and Pictured Rocks in the Upper Peninsula. And in fall 2019, six months before COVID hit, I led a nature writing workshop at a birding festival in Midland, Michigan. Midland was just down the road from Lake Huron, where we took a schooner ride out into the lake.

Summer 2022 brought us back to Michigan in a more sweeping journey. After visiting our daughter in Chicago, we swung around Lake Michigan's southern flank and began our Michigan tour at Sleeping Bear Dunes National Lakeshore, passed up through the Straits of Mackinac, and pressed onward into the Upper Peninsula through Marquette and the Keweenaw Peninsula. All told, we drove over one thousand miles of Michigan coastland.

Including its islands, the state of Michigan has over three thousand miles of shoreline along Lakes Superior, Huron, Michigan, and Erie, giving Michigan more shoreline than any other state, oceanside ones included. (T-shirt in a gift shop: "Four out of five Great Lakes prefer Michigan.")

Altogether, the Great Lakes comprise the largest interconnected freshwater system in the world, harboring 20 percent of the world's total freshwater supply. Lake Superior is the world's largest single freshwater lake by surface area, covering thirty-two thousand square miles. Spread out, Superior could cover the entire continental United States with a foot of water.

Along the Lake Michigan dune shoreline, at the Straits of Mackinac, on Mackinac Island, and even along the Lake Superior shore, I took any opportunity I could to wade in the Great Lakes waters. The average water temperature of Lake Superior in July is about fifty degrees Fahrenheit, causing it to bite around the ankles and shins.

The hearty take the full plunge.

## Wind

At Sleeping Bear Dunes National Lakeshore, we laid our bikes aside to climb a sandhill because, well, if there's a sandhill, you have to climb it. We left our sandals behind next to the bikes. You have to climb a dune in bare feet. These are rules, or so I opined to Dianne.

The heat of each sunbaked step on the soles of our feet turned quickly into relief as our heels and toes dug down into the cooler sand just inches below the surface. Halfway up, the slope steepened, and that drew our arms and hands into service as well. We joined the legion of four-legged, hunched-over human crabs scaling up the dunes.

Better to climb here on this hundred-foot-tall inland dune than on the shoreline where the dunes rise 450 feet from Lake Michigan. Signs at the top of the Shauger Hill overlook warn the curious that while the slide down to the lake may be quick, the climb back up can take upward of two hours, and the rescue charges are steeper than the dunes themselves.

The immense sand dunes at Sleeping Bear—at the shoreline and for a distance inland—are the product of ice, water, and wind. The main force reshaping the dunes today is wind.

Sleeping Bear Dunes National Lakeshore is a relatively recent addition to the ranks of national parks, with an Act of Congress creating it from private holdings in 1970. It stretches for thirty-five miles along the Lake Michigan shore, with interior dunes and forests bringing its total size to 111 square miles.

The park takes its name from the Sleeping Bear dune near its northern edge. Ojibwe lore tells of Mother Bear and her two cubs fleeing a massive forest fire (another version says they were fleeing warring tribes) on the western shore of Lake Michigan. Mother Bear and the cubs swam across the entire lake, seeking refuge on the eastern shore. They tired and weakened as they neared today's Michigan coast. Mother Bear reached the shore and lay down to rest, awaiting her cubs who were not far behind. But the two cubs foundered just offshore. Losing their will and their strength, they drowned. Their bodies became North and South Manitou Islands, visible just offshore. Mother Bear, never awakening, became the Sleeping Bear dune. While the Ojibwe had long known of the dune's peculiar shape, in

1721, the Jesuit explorer Pierre François Xavier de Charlevoix described it in his journal as looking like an animal at rest. He wrote, "The French call it *L'ours qui dort* (the sleeping bear)."[2]

The geological story of the dunes, on the other hand, begins with waves of glaciers creeping down from the north. The sand itself originated from the glaciers grinding up northern rocks and transporting the debris—boulders and ground-up sand alike—as the ice moved southward. When the glacier reached its farthest advance at today's Lake Michigan and began to melt, it dropped its load of sand into the newly forming glacial lake. When the low-water period set in, the sand became exposed on the dried-up shorelines.

Wind took the next turn in shaping the dunes as the lake levels rebounded. Relentless winds blowing across the refilled lake piled the sand up in steep-walled dunes against the eastern shoreline and blew it inland as well.

Wind continues reshaping today's shoreline dunes and inland sandy soils. The Cottonwood interpretive hiking trail at Sleeping Bear Dunes revealed cyclical successions in this wind-generated, sand-based landscape. Strong winds still pile up sand onto the dunes and blow loose sand inland from the dune tops. Where protected from the wind, sand-friendly grasses and shrubs like marram grass, sand cherry, and juniper take root and stabilize the sand. From there, the sand-flow chart can take two directions. On our hike, we encountered the first possibility, sandy sections amid the brushy plants, called blowouts, where the fledgling plant life has died out due to drought or disturbance. Removed of its cover, the sand goes on the move again, bandied about by the wind. Human and natural disruptions on the Sleeping Bear dune have made her barely recognizable today, her sands shifting gradually to the northeast.

The second possibility is that a vegetated section may progress toward forest cover. Where the brushy plants hang on, they set down strong roots and leave an annual mat of plant matter decay, both of which help keep the sand in place. A thin soil develops, and a few sun-loving birch and aspen take hold, laying the groundwork for a later forest of oak, pine, basswood, maple, beech, and hemlock.

But sections of forest will still cycle back to sand. A massive tree fall may disrupt the thin soil and expose the underlying sand. The hole in the canopy will burn out the understory and allow sand to take over again. Or wind will blow sand from a fresh blowout in the dunes on the forest edge into the tree line, suffocating the trees' roots. We saw ghost forests on our hikes, where stands of tall, sun-bleached, debarked tree trunks reached skyward from a sandy base.

Five-hundred-foot sand cliffs drop precipitously into Lake Michigan from Michigan's Sleeping Bear Dunes National Lakeshore. Michigan has sixteen hundred miles of shoreline along four of the five Great Lakes.

The Sleeping Bear Heritage Trail is a twelve-mile-long, paved cycling route that runs mostly through the wooded interior. But even as we bicycled into the thick woods a mile or so inland, we were occasionally reminded that these were not the woodlands of home where the dark soil goes on forever. As we sped along a downward slope of trail, I spied from the corner of my eye a rain-swept gully with its thin topsoil washed away and its underlying fine, yellow sand once again exposed.

Should wind find a pathway into the woods, the cycle will repeat.

Dianne and I ended our bike ride on the Sleeping Bear Heritage Trail with ice cream at the Shipwreck Café in Empire, MI. The couples and young families sitting near us at the long counter didn't notice the grisly tale of maritime disasters embossed beneath the counter's vinyl surface, hidden in plain sight.

There was the 1912 *Rouse Simmons* that sank in a November storm while carrying fifty-two hundred Christmas trees from Manistique, MI, en route to Chicago. Weighed down by the trees, the ship foundered overnight as the storm brewed. Two crewmen sent to check flashings were swept overboard. By the next afternoon, the ship sat low in the water as rescuers spotted it on the horizon. It disappeared before rescuers could arrive.

There was the 1958 *Bradley* that was almost safely home to dock when a call came in for it to return to its loading port to add to its cargo of quarried stone. Running late against an impending storm, it encountered 65 mph November winds that dashed its hull into rocks off Gull Island, splitting the *Bradley* in two and causing the ship's engines to explode. The *Bradley* had been scheduled to receive a new hull at its overwinter port had it concluded its trip safely. Only two of its thirty-five crewmen survived.

Within view of the Sleeping Bear Dune lay at least sixteen more shipwrecks. Nearly fifteen hundred wrecks from 1800 onward lay strewn across Lake Michigan. Estimates suggest that at least six thousand wrecks have littered the Great Lakes since the seventeenth century.[3]

Many of the Great Lakes maritime disasters have been caused by winds that grew wrathful over the waters where, as sailors have sardonically warned each other for centuries, the nearest land is straight down. Gordon Lightfoot famously wrote of the wreck of the *Edmund Fitzgerald* on Lake Superior, "That good ship and true was a bone to be chewed / When the gales of November came early." The November winds took the lives of all twenty-nine *Edmund Fitzgerald* crew members.

Great Lakes winds build up dunes, tear them down, bring rain and massive snowfall to their leeward sides, and give birth to the waves that take down ships.

But we don't speak of that in the Shipwreck Café, where ice cream melts down the sides of our cones and drips onto the sunken tales.

I awaken in the middle of the night to feel the tent rustling in the wind. Finding my sandals, a ball cap, and a long-sleeved shirt to slip over my t-shirt to guard against the chill, I unzip the screen fly and wobble upright outside the tent. We are in a tents-only section of a campground near the town of Empire, not far from Sleeping Bear, with tall eastern pines lining the site and separating us from our neighboring campers.

A moonless night, dark as could be, it is perfect for stargazing, at least where the stars poke through the canopy. It is so dark I don't actually see the pine branches above me. Directly overhead there are stars, numerous and brilliant, but then the stars simply stop well before the horizon. My brain tells me the pine branches obscure them, but my eyes tell me that the stars have simply petered out at the sky's edges.

When the wind lightly shakes the overhead branches, the black edges of sky expand and retreat with the sway.

I get slightly dizzy in the breeze and under the pulsing sky. I widen my leg stance to keep my balance and try to will the night back into solid form.

## Earth

After we loaded our bikes onto the Shepler Ferry that slid soundlessly out of port in Mackinaw City and under the expansive Mackinac Bridge, our first sight on Mackinac Island was throngs of bicycles awaiting rental.[4] Bikes of every size, shape, and color appeared to have been rounded up from random garage sales and pressed into service.

Bicyclists pedaled along the island town's coastal road, stopping to dash into or linger in a tourist shop, to snap a photo along Lake Huron, or to eat lunch at one of the numerous cafés. There were mountain bikes, road bikes, and tagalongs for kids. And those not partial to bicycles rode horseback or in horse-drawn carriages.

Even the shopkeepers moved about by bicycle or horse-drawn wagons. One worker pulled a ladder behind his bike on a makeshift trailer. Another used a horse-drawn cart to bring supplies from the port to their shop.

Automobiles have been banned on Mackinac Island since 1901, when residents noticed how much they scared the horses. Today, an auto-less culture has become the trademark of this four-square-mile, turtle-shaped island.

Dianne and I circled the eight-mile road's circumference twice, nodding to Lake Huron on the south, east, and north, and eyeing Lake Michigan on the western horizon. We rode up the short, steep incline onto the turtle's back, away from the port crowds and shops, zipping past the early 1800s fort and disappearing into remote forest roads. We rode past the ghost of an 1814 battle site and stopped at the community cemetery where we found the earliest grave dating back to 1849.

Mackinac Island may be a tiny spot of land surrounded by water at the joining of Lakes Michigan and Huron, but for the Anishinabek peoples—the Potawatomie, Ottawa, and Ojibwe of the Straits of Mackinac—this island was the center of the world. The island's full Ottawa name, *Michilimackinac,* means "Great Turtle," and tradition posits the island emerging on the back of a great turtle at the end of a flood that covered the earth. A sacred, life-giving place, it was also a place of Anishinabek burials dating back one thousand years.

Indigenous peoples have inhabited the Straits of Mackinac for more than nine thousand years. As with much of the Midwest, the beginning of human occupation is tied to the receding glaciers. During the low-water period of Great Lakes formation, the Straits were a mere river winding through the valleys. The earliest peoples may have lived in this bottomland near the river—artifacts of their existence were flooded by the developing lakes and are just now being discovered.

The lands of this watery region—the island and the protruding tips of the Upper and Lower Peninsulas—eventually enticed Europeans drawn initially by the fur trade and the desire to Christianize the Anishinabek. The first Frenchman to visit Mackinac Island was Jean Nicolet in 1634, while he was exploring the northern Great Lakes. French fur traders soon followed. As French interest in the region grew, Jesuit priests Claude Dablon and Jacques Marquette arrived in 1670, first establishing a mission on the island and then moving it to the north shore—the Upper Peninsula—in today's town of St. Ignace. Marquette had studied regional indigenous languages prior to his arrival, helping him to connect with the people.

In his writings, Marquette depicts an amicable relationship between himself and the Huron, another tribe of the region. Even when he was away, according to Marquette, "The Hurons came to the chapel during my absence, as assiduously as if I had been there, and the girls sang the hymns that they knew. They counted the days that passed after my departure, and continually asked when I was to return."[5] Marquette's journal describes Huron life in generally positive terms. Viewing the dances associated with the Festival of Squashes, he "did not find any harm" in them except for the "bear dance" wherein the women "dressed, growled, ate, and acted like bears."[6]

Seeking to expand their New World Territory, the French government soon sent Louis Jolliet to meet Marquette at St. Ignace, and the two set off on a new adventure. An hour north of my own home on the Mississippi River, I have often looked out from Iowa's Pikes Peak State Park to the confluence with the Wisconsin River. When Marquette and Jolliet entered here in June 1673, they were the first Europeans to sail on the Upper Mississippi.

But Marquette's heart must have been in St. Ignace because, when he completed his Mississippi River mission near Arkansas and established a mission among the Kaskaskia tribe in Illinois, he became ill, knew he was dying, and resolved to return to the Straits of Mackinac. He died during his return, on May 18, 1675, and his companions on the trip buried him along the way. Two years later, others disinterred his remains and transported them back to St. Ignace.

Dianne and I came upon Marquette's grave fortuitously. Upon leaving Mackinaw City and driving across the Mackinac Bridge, on our way (ironically) to Marquette, MI, Dianne discovered the Museum of Ojibwa Culture in St. Ignace, and we stopped there before the next leg of our journey. Before entering the building, we spied a white, weathered cross monument in the courtyard, its base bearing a plaque proclaiming: "In memory of Rev. Father James Marquette . . . who died at the age of thirty-eight and was buried in this grave AD MDCLXXVII [1677]."

The museum tells a nuanced story. The opening displays recount major wars and battles Native Americans have fought against French, British, and American forces, yet also detail how indigenous peoples have fought in subsequent American wars. Largely celebratory regarding the legacy of Jacques Marquette as one who intuited that the Anishinabek already knew their creator, they also speak of other Jesuits who belittled indigenous peoples, of the horror and indignation of Christian-run Indian schools, and of the sordid history of "Indian Removal."

Some Euro-Americans viewed this struggle as a battle for souls, but for most it was a battle for land.

---

European and American military forces followed soon after the Marquette era.

The French built Fort de Buade at St. Ignace in 1683, but soon replaced it with Fort Michilimackinac at present-day Mackinaw City on the lower peninsula in 1715. Dianne and I toured the reconstructed fort, climbing onto the cannon-equipped corner guardhouses, ducking in and out of the restored wooden cabins of the guardhouse, soldiers' barracks, and priest's house. The historic site's employees were dressed in period clothing, telling us about their annual fur-trading journeys and the officers' wives challenges of cooking on a fireplace. Occasionally they slipped out of character to talk with us about last winter's weather. When we descended into a belowground museum, we overheard two employees on break, still costumed, talking excitedly about recent archaeological finds.

The reconstruction has been true to the site's archaeology. Digs were still in progress at the fort, and we talked with an archaeologist who was currently unearthing a trader's house on the grounds. Stoneware sherds, brooches, buttons, shoe buckles, and cufflinks suggested that this was not a soldiers' quarters. When the dig was completed, she said, the cabin would be reconstructed as it would have stood on the grounds.

The British replaced the French at Fort Michilimackinac in 1761, a year after defeating the French at Quebec. During 1780–1781, the British moved the fort to Mackinac Island for a more sweeping view of the Straits. They burned the abandoned fort at Mackinaw City. Sand swept over the charred remains, providing future archaeologists with a starting point to reconstruct Michilimackinac.

By 1796—thirteen years after becoming American territory—the British ceded Fort Mackinac on Mackinac Island to the American army. But the British captured the fort again during the War of 1812 and did not return it to American forces until 1815. In the ensuing decades, the US Army repeatedly abandoned the fort to allow its troops to fight in the Mexican War, Civil War, and other military ventures, and the returning troops would reoccupy it in the aftermaths.

In 1875, the fort's mission changed entirely. Fort Mackinac became army headquarters for administrating the newly minted Mackinac National Park. But in 1895, the army decided to get out of the national park business and turned the island and decommissioned fort over to the state of Michigan.

Peering out from a raised guardhouse cannon slit at Fort Michilimackinac, I could see the Mackinac Bridge linking the Lower and Upper Peninsulas of Michigan across the Straits, and linking the eighteenth and twenty-first centuries in one view.

Two days later we crossed the five-mile-long bridge and headed to Marquette on the Upper Peninsula.

## Fire

Properly helmeted and after a brief, explanatory film, Dianne and I descended 110 feet into the uppermost level of the long-defunct 1847 Delaware Copper Mine. Alongside the stairway ran old rails that, more than a century ago, hoisted copper and waste rock up to daylight.

Seepage from recent rains glistened, dripped, and trickled into the descending shaft, then disappeared into dark passageways. Electric lights encased in faux lanterns lit the lateral vein, but they were soft, subdued, and spread far apart along the passageway. They gave off a light as wan and copper-colored as the hard-sought ore, offering just enough light to peer into subterranean rooms and into cross-cut passageways. The tops of shafts descending to the lower levels were barely visible, their shadows occasionally revealing the water surface of the seepage that had filled the several lower levels—the deepest shaft descending fourteen hundred feet—in the years following the mine's closure in 1887. The topmost level was reasonably dry, as it drained through a distant adit, or horizontal vein, that opened to the outside world.

Inside, the temperature was a constant forty-three degrees Fahrenheit. A white fungus grew on some of the wooden structures, the result of the damp air. Timbers still supported the bedrock roof, but bits of discarded timber and rail lay along the passage. A few supplies had been left behind in the mine as well, including a long, hand-cranked drill used for boring into the rock.

The tour was self-guided, with occasional explanatory signs along the passage. Dianne moved down the passage more quickly than I and eventually disappeared from view. I was alone, then, 110 feet below ground.

In 2012, scuba divers explored the flooded lower levels of the Delaware Mine. Their YouTube video shows even more old rails, support timbers, abandoned buckets, ore cars, ladders, and tons of rubble submerged in the water throughout the passageways. In one section, there had been a roof collapse, but not enough to block the passage.[7]

Dark and damp, cold and flooded, is the world deep inside the mine. Yet ironically, the story of the Upper Peninsula begins in Fire.

---

Sleepy sedimentary limestones, shales, and sandstones underlie much of the Midwest. This bedrock formed in shallow, equatorial waters hundreds of millions of years ago before the North American continental plate drifted northward and uplifted. In the eastern Upper Peninsula of Michigan, iron-rich sediments on the ocean floor were later transformed into iron ore by the extreme heat and pressure of newer layers of sediment laid down on top.

By contrast, the western half of Michigan's Upper Peninsula bedrock was birthed in fire from the start. These are igneous rocks—granites and black basalts formed 3.5 billion years ago—and metamorphic gneiss, schist, and quartzite transformed by heat and pressure applied to granite, shale, and sandstone, respectively.

A midcontinent tear in the tectonic plate, sometimes called the Keweenawan Rift, began to stretch and thin in the Upper Peninsula 1.5 billion years ago and continued to split down-continent. North America came close to splitting, but it didn't. Even so, molten rock burbled up into the rupture, filling in the underground gaps as magma and occasionally spilling forth as lava.

Copper was formed during this period and in this same manner, with its molten mineral base gurgling up into the rift fissures during massive episodes of a kind of continental acid reflux.

Fire, fire.

The Iron Ore Heritage Trail that logs forty-seven miles across the Marquette Iron Range on former train beds in the Upper Peninsula eases into the city of Marquette's southern flank. We picked up the trail on the southern shore of Lake Superior. The trail soon meshes with Marquette's city trails—one of which connected to our campground—and leads to Presque Isle Park on the city's northern side.

The Iron Ore Heritage Trail is not just named for the past. While stopping for lunch at Presque Isle Park, we watched railcars edging onto the iron ore dock to drop iron pellets from regional mines into the two hundred "pockets" that will eventually fill ore ships headed for steel mills 150 miles away across Lake Superior at Sault Ste. Marie. Almost 10 million tons of ore are still shipped annually from this dock.[8]

Land surveyor William Austin Burt discovered iron in the Upper Peninsula in 1844 when his compass needle began behaving erratically near Teal Lake, twelve miles from Marquette. Burt called out to his workers, "Boys, look around and see what you can find!" They soon found chunks of iron protruding from the hillsides.[9] An iron mining rush soon commenced, supplanting iron extraction in other parts of the country because of the ease of its removal and transportation. By World War II, 103 iron mines were operating in Michigan.[10] Vastly fewer mines operate today, and all are open-pit operations, but Upper Peninsula mines still supply the bulk of the nation's iron ore production.

But the red ore alone didn't hold my attention for long. I was equally interested in the black rock 150 feet beneath the pines of Presque Isle Park. The chocolaty rock formation is a hard layer of metamorphosed igneous rock that juts out at the base of a softer, rusty orange sandstone that arises from Lake Superior. Several paths lead down from the cliff top to the black rocks, which lie about ten to twenty feet above the lake. And because

Superior is about twenty feet deep most places at the edge of the black rock, cliff jumping is a favorite local activity. Not previously knowing this, I did a double take when a young man suddenly sprinted toward the ledge, flipped in the air, and disappeared into Lake Superior, all without drawing any particular attention from others combing over the rocks around him.

Rock of another color—pink, glacially scratched gneiss and granite at the southern edge of the Canadian Shield—rests atop 470-foot Sugarloaf Mountain about six miles south of Marquette. The granite is 2.7 billion years old, among the oldest bedrock in North America. The path to the peak is moderate, varying from dirt paths that share their space with gnarled cedar roots, to stepping-stone stairs chiseled or wedged into the bedrock, to wooden stairways leading over the tallest of the weathered outcrops.

Some chose the more challenging way upward, clambering over the rocks like goat kids. They were human kids, though. Dianne and I were glad to see their parents letting them scramble along and atop the rocks without admonition and without ordering them back to the path. "I'm going to become a rock climber," one boy called back to his dad. "Rock climbing is hard, and this is hard!" His dad, of course, was perched precariously at the edge of a drop-off, so it probably ran in the family's genes.

The view from the peak sweeps across Lake Superior to the north and to a northern boreal forest of aspen and maple to the west. The soft-edged forest range before us belied a bedrock born of fire.

Several years ago, Dianne and I visited the Upper Peninsula's Porcupine Mountains.

Known colloquially as the Porkies, the range rises one thousand to sixteen hundred feet above Lake Superior, forming parallel ridges a mile or so inland. Glacial lakes, marshes, and old-growth virgin forest separate the parallel ridges. From a fire tower at the crest of the Summit Trail, the highest point in the Porkies, the ridges undulate from peak to peak, forming a silhouette against the horizon that reminded early inhabitants of the hunched-up back of a porcupine.

We spent most of our hiking time on the Escarpment Trail, admiring sheer basalt cliffs that drop precipitously into the Lake of the Clouds five hundred feet below. The sky feels close here and is doubled in size by its reflection in the lake.

The escarpment—the term means a long, steep-sloped ridge—was likewise formed during the Midcontinent Rift. Volcanic magma rushed in to fill the void along the tear, producing a hard igneous rock that persisted even as later glaciers slid over it. Hiking along the ridge, we could still see scratch marks left on the exposed bedrock by the grinding ice.

The Porkies are home to over thirty thousand acres of old-growth virgin forest, including the largest stand of hardwoods west of the Adirondacks. Towering eastern hemlocks—hundred-foot-tall, short-needled pines—shade large sections of the forest floor, preventing other trees from getting established and resulting in a "clean" understory from which the hemlock trunks arise like cathedral columns. Large swaths along Lake Superior were logged until the remaining old-growth Porkies forest was declared a wilderness park in 1945.

A short-lived copper mining rush blew through the Porkies like a wildfire in the 1840s. We hiked the Union Mine Trail through the bottomlands where interpretive signs point out fading evidence of an 1846 mining community. In October 1846, miner William Spaulding wrote home that "everything seems to prosper at this location; we are not only turning out copper, but children. Mrs. Shin gave birth to a whopping baby boy today."

The mining boom ended quickly. The tone of Spaulding's February 1847 message home is gloomier, almost foreshadowing the demise of the mining community: "A man by the name of Baily was found frozen to death four miles from the mouth of the Iron River." They abandoned the Union Mine later in 1847 due to slumping copper prices. It was reopened briefly in the 1860s and again in the early 1900s, but each time closed quickly due to the vagaries of the market and limited ore deposits.

A short walking path outside of the Delaware Copper Mine site on the Keweenaw Peninsula takes us past the ruins of the operation that ran from 1846 to 1887. Tall white-trunked birches have softened the memory of the roof-caved powerhouse, hoist house, and engine house.

Glass-free window openings in the stone facades reveal the forest taking shape within the structures. At the edge of the parking lot sits a cascading slide of waste rock, since the mines typically only yielded about 2 percent copper. While mounds of waste rock still dot the countryside, some communities have belatedly given a portion of this rock a second life, further crushing it for use as road or shoulder gravel.

A path in the other direction takes us down to a five-thousand-year-old, prehistoric Native American copper mine on the Delaware site. Indigenous peoples had long ago exploited the fissure in the bedrock here to find a copper vein. Using fire to heat the rock and water to split it open and separate the copper, they then used stone hammers to shape the copper into tools and artwork. Copper work from Michigan's prehistoric mines has been found at archaeological sites across North America—the evidence of vast, ancient trade networks.

By the 1670s, French Jesuits and explorers were sending home reports of copper lodes found or rumored to be in the Keweenaw Peninsula. French efforts to establish profitable copper mines in the area were "sporadic but unsuccessful," according to mining historian David J. Krause, PhD.[11] Under brief British control between 1763 and 1783, speculators continued making sporadic mining attempts, but potential investors back home doubted whether copper mining could be profitable unless gold were discovered alongside the copper.[12]

The American copper boom in the Keweenaw Peninsula began in the early 1840s. According to Krause, "In 1840 the south shore of the lake was still mostly a wilderness," but by 1850, "towns had been established at Copper Harbor and Ontonagon. Roads were being built, [and] local trade was beginning."[13] Ninety-one companies tried their hand at mining before 1865. Most of the new mines were not successful, with eighty-six of the companies paying no return on investment whatsoever.[14] The Delaware mine lasted longer than many, but after thirty years, it had not yet returned a profit. The Cliff mine near the now-abandoned town of Clifton was the first to strike veins large enough to be profitable, and a few others followed.[15]

Annual production reached its peak in 1916, with nearly 270 million pounds of copper mined from the region.[16] Copper mines throughout the peninsula shipped their ore through Copper Harbor, giving it a bustling population of thirteen hundred by 1887.

A slow decline set in after the 1916 peak, and by the 1960s, most of the Keweenaw Peninsula copper mines had closed. Left behind were crumbled economies that led to a decreased population. Keweenaw County's 2020 population of 2,046 was only 29 percent of its high mark of 7,156 in 1910. Between 1940 and 1950 alone, its population declined by 25 percent.

Thus began a gradual shift to tourism as the Keweenaw Peninsula's major economic focus. Today, tourism accounts for 70 percent of the county's workforce.[17]

Our campsite at the Hancock Recreation Area at the base of the Keweenaw Peninsula sat alongside the Portage Canal that actually turns the peninsula into an island and offers a shortcut along Lake Superior's southern shore. In its natural state, the peninsula was connected to the mainland by a narrow isthmus. Native Americans and early European explorers could portage the isthmus. Canal construction began in the 1860s. Today, its twenty-five-foot depth and a lift bridge between the towns of Houghton and Hancock allow vessels of significant size to use the shortcut.

We were interested in no shortcuts that afternoon. Dianne and I are often on the move on trips like these. We see a lot of places and talk with a lot of people. On this day, though, at the end of a long drive, we set up our tent and parked our camp chairs next to the canal from late afternoon onward, sitting, reading, and thinking at the edge of the shimmering water.

Let this fire go down to embers.

On our route north through the Keweenaw Peninsula, the road sign announced the town of Calumet, ten miles ahead. Calumet, Michigan, where did I know that name from? "Take a trip with me in 1913, to Calumet, Michigan, in the copper country." It was a song, but what song?

"I'll take you to a place called Italian Hall, where the miners are having their big Christmas ball."

It came back to me, line by line. It was Arlo Guthrie's voice, singing his father Woody Guthrie's song, "1913 Massacre."

We have to turn into town, I told Dianne.

The Keweenaw Peninsula copper mines had seen labor strikes in 1872, 1874, 1890, 1893, and 1906. Wages remained low, but by 1910, mine owners were introducing new machinery to save labor costs, including lighter-weight drills that required just one man to operate, not two. For the miners, though, fewer persons in the shafts not only meant fewer jobs, but also increased safety concerns with fewer coworkers watching out for each other. By summer 1913, Keweenaw Peninsula miners had had enough and went on strike at the Calumet mine. In addition to their safety concerns, miners pressed for higher wages and an eight-hour workday.

The strike dragged on through the fall and winter. Violence broke out immediately. The mine owners soon hired strikebreakers who shot and killed two striking miners for trespassing. Refusing to negotiate with the strikers' union, the owners hired "scabs" to reopen the mines.

By Christmas, the strike was in its fifth month and miners were running out of resources. To raise spirits, the Western Federation of Miners' Ladies Auxiliary organized a Christmas Eve party for the strikers and their families in the upper floor of Italian Hall, a local gathering space built by and for immigrant mine workers.

While children were receiving Christmas presents upstairs, someone outside the hall yelled "Fire!" A panic ensued, with parents and their children dashing down the stairs to a door that strike-breakers and hired thugs had barricaded. Seventy-three persons—more than half of them children—were smothered on the stairs by the crush of people trying to get to the door. Despite an investigation, no one was ever arrested.

Dianne and I drove the streets of Calumet (population 776), looking for the site, which had been commemorated by the town in 1989.

Finally, we found it. Although Italian Hall had long ago been razed, the door's fateful arch had been preserved. Behind it, visible through the archway, sits a granite monument emblazoned with the names and ages of all seventy-three victims. The youngest had been Rafael Lesar, age two.

The Calumet mine finally closed in 1968.

"See what your greed for money has done," the Guthrie song concludes.

Friends had told us, "We love Copper Harbor," as we planned our route through Michigan, so much so that I was bracing myself for a bustling tourist town. I envisioned circling the streets to find a parking spot, crowded sidewalks, and waiting lists at restaurants. Our friends must love uncrowded places, though, because as we descended the long hill into Copper Harbor on a Thursday morning in July, we found a town going about its own business. There were motels and bed-and-breakfasts for visitors, but they seemed modest enough in size and décor, and all locally owned. We pulled our bikes off the car rack at the city park, parked alongside a handful of other visitors, and cycled easily enough around town. We found a hidden bookstore named Grandpa's Barn—its contents rivaling those in the cities—where I bought some books on local copper mining history.

A small visitors' center told the story of mining throughout the peninsula, and of Copper Harbor as its principal shipping port during the short-lived rush. Today, only around one hundred people live in Copper Harbor year-round, and about three hundred in summer. There is no Wi-Fi, and the town's dozen or so kids still attend a one-room schoolhouse.

Copper Harbor remains unincorporated. One local explained, "We don't have a lot of rules, and we like it that way." Then he paused and added, "We have 'covenants' instead," or expectations of how to behave and to help each other as needed. Townspeople use a message board to offer rides or to pick up supplies when they drive to distant towns.

We bicycled a bit farther down the road toward the final tip of the peninsula and Ft. Wilkins State Park. The military post was established in

Copper Harbor, at the northern tip of Michigan's Keweenaw Peninsula on Lake Superior, boomed in the mid-1800s as a shipping harbor for copper mined from the region.

1844 as the copper rush ramped up but was abandoned by 1870. We cycled around the campgrounds and park roads, stopping at the shore of Lake Fanny Hooe at the edge of Superior. Fanny Hooe has a deep, barrier-free connection to Lake Superior that creates two thermal layers, a sun-warmed upper story for bass, perch, and pumpkinseed, and a cold-water basement for brook trout and occasional salmon visitors from the Great Lake.

Hannah Rooks's father always told his clients to listen for the "quiet full of noise" in the Northwoods. He wanted them to listen for birdsong and the rustling of leaves, she said.[18] Her father, Jim Rooks, ran Bear Track Eco Tours in Copper Harbor, MI. Before that, he had been the first director of the E. B. Lyons Preserve at my home in Dubuque, IA, and the city's naturalist from 1974 to 1983. Born in Michigan, he felt the call to return, not just to the state, but to one of its most remote locations.

On the Brockway Mountain Drive above Copper Harbor lies the James Dorian Rooks Nature Sanctuary. It had been a quest of mine to find the site ever since we'd made plans to visit the Keweenaw Peninsula.

The sanctuary includes a 1.2-mile hiking trail that loops through a well-shaded woods with occasional overlooks of Lake Fanny Hooe and Lake Superior. Harsh winds from Superior create a microclimate here, favoring trees and wildflowers more prevalent in Canadian forests. In the valleys the oaks grow tall and sturdy, but on the exposed ridges they are small and stunted, like young oaks captured in old gnarled bodies. The understory is rich with ferns. On this July day, orange fox-and-cub wildflowers showed off against the white trunks of downed birches.

It was a lovely hike. But it wasn't the sanctuary per se that I had come to see. It was the fact that the refuge was dedicated to Jim Rooks.

I met Jim Rooks only briefly around 1980, when I wanted to write an article about the new preserve for my college newspaper in Dubuque. I arranged to interview Mr. Rooks at the Lyons Nature Center. Grabbing a notebook, I anticipated sitting in his office, asking him a few questions, and going home to write an article. Jim had other ideas.

"I've got an errand to run," he said as I arrived. He had a catfish from the basement aquarium that he wanted to release into the river. "Let's get in the truck."

We bounced out onto the road, Jim talking about the preserve and me taking notes that looked like a seismograph printout during an earthquake. Then he pulled off onto the shoulder, hopped out, and picked a dead butterfly off the pavement. He was building a display about monarchs back at the preserve, and he'd make use of this specimen. He returned to the road but veered off again at his roadside mailbox. Grabbing a stack of bills and junk mail, he lamented, "No one writes letters anymore. I like to write." At another juncture, he pulled into a driveway to show me a stone wall whose mason must have worked about the same time as the 1840s limestone structures on the Lyons Preserve.

It took me a while to realize that the journey was part of the story.

Forty years later, we found the Laughing Loon, the gift shop Rooks and his wife Laurel opened when they moved to Copper Harbor, and that Hannah now runs since her father's death and her mother's retirement.

"He had a great deep voice," Hannah says of her father, and it helped him draw people into the wonders of the natural world through his story telling and passion. And he had a bit of a devil-may-care attitude: "The road's all rock," he'd tell a customer worried about how Rooks's front-wheel-drive van would fare in the Copper Harbor woods. "We're not gonna get stuck!"[19]

Locals would call Rooks when they spotted an injured animal. "We always had a recovering animal in the house," Hannah recalls, "an owl, songbirds, an injured duck taking up the shower." Now that her father had passed away, locals call Hannah when they find an injured animal, or she discovers them on her own. "Yesterday I drove a little pink mouse I'd found on my driveway forty-five minutes to the rehab center."

Rooks was active in Copper Harbor conservation. In the 1980s, he stopped loggers from encroaching on the protected Estivant Pines, a stand of three-hundred-year-old white pines that reach 125 feet skyward near Copper Harbor. He also helped to procure Hunter's Point Park, a thin peninsula that creates the town's harbor. Hunter's Point may well be the northernmost peninsula of the northernmost peninsula of Michigan's Upper Peninsula.

When Rooks passed away in 2005, the Michigan Nature Association named the Brockway Mountain nature sanctuary in his honor.

Jim Rooks used to walk the trails at this sanctuary. And Hannah was going there after we talked.

This was another kind of fire, the kind that came from within and ignited others.

## Four Elements

On the last morning, we reloaded the Honda and recrossed the Portage Canal Lift Bridge in Houghton, then headed south of the Keweenaw Peninsula, with our home in Dubuque as the day's destination.

Water disappeared first, as Lake Superior diminished in the rearview mirror.

Then we crossed over the earthy hump of the St. Lawrence Divide and began a slow, southern descent across the watershed.

Sun fire disappeared next, extinguishing behind an ominous cloudbank.

A last blast of wind propelled a summer squall, and then it was gone, too.

Four hundred miles put Michigan behind us, and Wisconsin as well. The bridge that crosses over into Dubuque slings down from a limestone bluff. The sun had returned, and beneath us was water again: the glistening Mississippi rippling in a gentle breeze.

# 3

# Ohio

## *The Hocking Hills: What Lies Buried*

I wished I owned windows like these. Light streamed in from seven Gothic openings stretching almost from the floor to the twenty-five-foot-high ceilings. The openings were irregular, curved, carved, twisted, and phantasmagoric: an ace of spades, a bowling pin, a goose taking flight, tongues of fire. From the outside they could barely be seen, but from the inside the seven windows cast light spells across the otherwise darkened room.

Dianne and I were in the Rock House cave in southern Ohio's Hocking Hills. To get to the cave, we'd dropped halfway down into a two-hundred-foot valley on a trail that twists from the parking lot. Largely hidden within a massive sandstone wall, the cave is accessible only through its seven windows. This cave—a luxurious space running two hundred feet long, twenty to thirty feet deep, and twenty-five feet tall with interior recesses for cooking fires and baking—has been home to ancient peoples for thousands of years. The windows are water- and wind-hewn openings to the cave's interior.

We and a few other families were merely passing through, of course. We used our cellphone flashlights to negotiate the deeper recesses and uneven floors where the light didn't penetrate. We looked out over hills stretching beyond the windows. We basked in the coolness of the chamber, and in the perfect balance of light and earth and underworld.

Were it not for modern signage, the cave would be easy to miss, which, I suppose, made it an even more prized ancient dwelling. Its hidden nature also gives weight to the mystery of what lies beneath and behind the surface in southern Ohio. Is there treasure or tragedy behind and beneath these openings in the rock?

---

Dianne and I had entered Ohio from Indiana through a more common doorway on Interstate 70, midway along their shared border, and before long found ourselves Google-Maps-lost amid construction in Dayton. At Columbus we hooked southeastward toward Lancaster, where we'd spend the night on the edge of the Hocking Hills. We pulled into Lancaster late in the afternoon to savor one last night in a real bed and a morning hotel shower prior to several nights of camping. But as we took an evening walk and went to dinner, the steep hills just south of town piqued our curiosity. The glaciated flatlands we'd passed through in central Ohio had ended, and now we were on the edge of deep rock.

---

We entered the hill country gradually. Just south of Lancaster the next morning, we drove to Clear Creek Metro Park. Two outcrops along the park road offered us glimpses of the underworld to come. Written Rock is a kaleidoscopic explosion of swirling yellows, grays, blacks, and reds. The reds hint at iron ore.[1]

What exactly is "written" in the rock I'm not quite sure, except for a small amount of graffiti, probably more recent than the rock's given name. I'd like to think that geologists were thinking poetically, as Written Rock foreshadows the story of the Black Hand sandstone that underlies the region's spectacular geology like a running theme. Black Hand sandstone runs 150 feet thick through the Hocking Hills, with a hardened topmost layer overlying softer layers that erode out from beneath when exposed.

Leaning Lena, the other bedrock intrusion along Clear Creek Road, is a breakaway slab of Black Hand that fell when the softer sandstone beneath it gave way. Lena lodged itself into the adjacent soil at a forty-five-degree angle, with just enough room for the road to squeeze between the fallen slab and its mother bluff.

The region, the county, the hills, the state park, and the state forest all take their name from the Hocking River. The name originates with the

Shawnee, who had called it the Hockhocking—or Bottleneck—due to a straight stretch of the river that suddenly broadens before a waterfall. The area's rugged terrain, as well as its remoteness and jaw-dropping cave and cliff features, saved it from the ravages of mining and logging for over a century. The state of Ohio began legal protection of the ninety-nine-thousand-acre Hocking State Forest and twenty-three-hundred-acre Hocking Hills State Park through land acquisitions in 1924.[2]

Water chased us down to the Rockbridge.

One of the northernmost features of the Hocking Hills region is the Rockbridge Nature Preserve. Rockbridge is Ohio's largest natural bridge, spanning over one hundred feet across a fifty-foot gorge on a hanging ledge five to twenty feet wide.

The forecast gave us no warnings, and the sky made no dire pronouncements as we set out on the grassy trail from the parking lot behind a small group of twenty-something couples. But a minute or so down the path, light sprinkles started falling, so Dianne hustled back to the car and grabbed some thin, plastic ponchos while I shot some photos. It was just a precaution. But two minutes down the trail again, the sprinkles turned into torrents, and we were soaked before we could don the ponchos, two thin, yellow, plastic, hooded coverings that made us look like giant canaries.

We soon caught up to the young couples, who were now huddled under a tree, soaked to the skin and laughing. They recognized us from the start of the hike, now draped in our waterproof Big Bird suits. They greeted us as people only do when mutually and ridiculously wet. I must have seemed particularly comical, with my yellow hood topped off by a dripping baseball cap.

They kept laughing as we passed. "I've never robbed anyone before, but . . . ," one of the young men said of my yellow suit.

The torrent followed us down into the forest. The canopy was still intercepting most of the downpour, but a small rivulet formed in the uplands and chased us down the trail and then hurried past. It was on its way to the Rockbridge too.

We descended gingerly along the wet trail. The forest was dark under the gray sky. Around a corner at the base of the trail, the Rockbridge came darkly into view, spanning the gorge that had come to life with runoff.

Here again the layering of Black Hand sandstone was apparent. Above the bridge, the local stream had nosed out a weakness in the surface and chiseled down into the bedrock, tunneling under the Black Hand cap. The upper portion of the cap had long ago collapsed into the cavity, but a five-foot-thick narrow band called the Rockbridge hung on and still spans the gully.[3]

Crossing the Rockbridge at its twenty-foot width was of no concern, but we gave the rock due regard where it narrowed to five feet. The wet sandstone was slippery beneath our sneakers, and the slight undulations on its surface might easily catch this stumble-prone hiker's stride. Sidestepping puddles, we paused as the creek passed beneath us, the underworld erupting onto the surface.

I wondered about the slow accumulation of sand on the sea bottom hardening into rock, the gradual uplift of land and the downcutting of the stream—all a slow and patient evolution emerging at the moment of *now*, as Dianne and I stood at the center of the Rockbridge.

Then we turned away and let another *now* take hold, one without us, one that involved the young couples who'd caught back up with us and were posing for photos on the slippery rocks.

---

Hocking Hills State Park contains seven distinct units linked by trails and roads: the Rock House Cave, Cantwell Cliffs, Ash Cave, Old Man's Cave, Whispering Cave, Cedar Falls, and Conkle's Hollow.

Later we'd find the park to be overflowing with summer visitors, but few people had found the Cantwell Cliffs, the northernmost and farthest removed attraction. The trail begins with the misogynistically-named Fat Woman's Squeeze, where I waited behind an older man who descended the hewn-out stairway in two-legged stutter steps. The narrow stairway passed between towering sandstone block walls en route to the valley floor.

At the cliff, a stream leaps from the resistant layer of Black Hand into the 150-foot canyon it has carved into the softer rock beneath. Reaching the base, Dianne and I clambered into the U-shaped rock shelter behind the falls.

Rock shelters are perhaps the signature feature of the Hocking Hills. Ash Cave is the most massive, with its horseshoe-shaped rock shelter stretching seven hundred feet from end to end and one hundred feet deep behind its ninety-foot waterfall.[4] It takes its name from deep spreads of fire pit ashes examined in an 1877 archaeological study. One three-foot-deep ash pile stretched for thirty feet. The ashes included arrows, animal bones, flints, and corncobs, indicating intensive use by Native Americans from

Southern Ohio's Hocking Hills host numerous caves, rock shelters, canyons, and waterfalls. Ash Cave stretches seven hundred feet from end to end, with an interior recess of one hundred feet tucked beneath the sandstone overhang.

the end of glacial times until Euro-American contact. Indigenous travelers had used the rock shelter as both a dwelling place and a temporary shelter along a trail that connected Shawnee villages.[5]

Ash Cave reminded me of the world that everywhere undergirds us. What vast foundations of rock do we stand on, what subterranean lakes and streams lie beneath us? How do we keep our balance atop such depths?

Perhaps we do it by clinging to a wall of sandstone in a rock shelter halfway between the underworld and the surface?

Or do we surface to live in the moment? And at that moment, I was watching children play in the splash of the waterfall.

I may have swished around at the pool's edge myself. Dianne knows that, at nearly any location of running water, my shoes and socks come off. I am a Labrador-Retriever-water-hound while she is a keep-me-dry cat.

Old Man's Cave was the busiest area we encountered at Hocking Hills, likely due to its proximity to the visitors' center. It takes its name from a nineteenth-century hermit who lived in the rock shelter recesses. Old

Man's Cave has less of a precipice and waterfall than some of the other sites, but here Queer Creek tumbles over two sets of falls that spill over layered ledges, empty into crowd-pleasing pools, and finally disappear into the Devil's Bathtub, a circular pothole long ago bored out by the swirling stream. Handsome Depression-era bridges constructed by the Civilian Conservation Corps (CCC) cross Queer Creek for scenic viewing along the half-mile gorge. One-way traffic signs direct the flow of hikers.

I preferred the quieter reaches of Whispering Cave, a long way from the state park crowd. Whispering Cave is likewise a horseshoe-shaped rock shelter cut from the Black Hand sandstone, but its water sources run only in winter melt or after long rains. With only a few travelers sharing the wide arc, we sat quietly among the cave's many ledges. I could imagine overwintering here.

I examined the honeycomb-weathered rock at the shelter's edges. We gazed down at the scattered rubble of breakaway boulders and sandstone slabs that littered the still-descending gorge, as Whispering Cave is nestled on a ledge above the valley floor. We talked with a fellow traveler, a young man who had recently moved to the area and was discovering it for the first time along with us.

Whispering Cave was so named because sound carries easily from one end of the cave to the other, but our talk was quiet simply because we seemed to be in the midst of a sacred place.

The hike back to the visitors' center was long. We leap-frogged alongside a younger couple for much of the way, as each of us stopped for photos or rested at alternating locations. Dianne drifted off to the side of the path to photograph wildflowers. We had been in the cave's darkness a short while earlier, so I reminded her that her camera flash was still on. "Yes," she replied, mishearing me, "you've still got your sunglasses on."

I caught a condescending smile pass between the younger couple as they overheard the miscue while passing us by. "What a cute old couple," it seemed to convey. And then we smoked past them back on the trail.

Hemlock trees define the last two units of the Hocking Hills State Park. Cedar Falls had been misnamed when early settlers thought the prevailing trees were cedars, not hemlocks. Here Queer Creek slides down a sandstone slope, separates into two flows around a boulder halfway down, regroups in a hollowed-out depression, and spills twenty-five feet into its basin pool. Although it has less of a freefall than many of the others, the water volume of Cedar Falls is the greatest of any in the park.

Conkle's Hollow makes for a good last hike if you've been walking all day, as a handicapped-accessible sidewalk stretches half the length of one of Ohio's deepest gorges. The valley floor abounds in ferns and wildflowers, and hemlocks grow so thick on the slopes and ridge tops that the sunlight here is tinted green. The hemlocks are a cold-climate relict species, more prominent in the area thousands of years ago when the glaciers had halted nearby. Today they are usually found in more northerly climates. This far south they thrive in the cool, moist hollows.

A trail continues up the valley after the sidewalk ends, clambering across slumped boulders in the ever-narrowing gorge until arriving at the Conkle's Hollow waterfall. The stream entering the gorge has already downcut halfway through the sandstone when it punches through a canyon wall and spills down into its base pool.

A more strenuous upland trail rims the gorge, but we decided to call it a day.

The Hocking Hills reminded us that our lives are lived in a particular time and place. We occupy a small space on the narrow mantle between the earth's core and the upper atmosphere. We live at the emerging tip of time pushing forward like a spring bud. But time is circular as well. This same space is layered with earlier lives. The present is simply fresh soil spread thinly over the old, all the way down to bedrock.

I had seen the bedrock sandstone. I also needed to see the lives and cultures that had undergirded this place.

We pulled into the visitors' center of the Hopewell Culture National Historical Park in Chillicothe, about an hour west of Lake Hope. My home in northeast Iowa, with its abundance of Native American mounds along the Mississippi River, owes a debt to the Hopewell culture that originated and flourished in southern Ohio from 200 BC to AD 400. Chillicothe was home to at least six different Hopewell sites with conical, linear, and elliptical burial mounds, as well as circular, rectangular, parallel, and perimeter earthen walls. Many of the sites include celestial alignments.

Park Ranger Jacob greeted us as we entered the visitors' center. We told him that our interest in the Hopewell mounds stemmed in part from the mounds in our corner of the upper Midwest. He multiplied our enthusiasm. "An indigenous renaissance took place here," he said with a sweep of the hand while offering that he himself was part Cherokee.

Hopewell culture spread throughout the Midwest two thousand years ago. But that doesn't mean the Ohioan peoples themselves had migrated across or conquered the Midwest, Jacob explained. That the Hopewell culture spread without the mass movement of peoples suggests that it was adopted through trade and cultural interaction rather than having been forcefully imposed through conquest. "The culture spread widely as people freely adopted Hopewell ideas," he said.

Mound building was central to Hopewell culture, but was not its only defining feature. Nor did it originate with the Hopewell culture. Six hundred years before the Hopewell, the area's Adena peoples were mound builders. The Hopewell culture intensified and perfected the Adena mound-building practices, and added their own distinctive traditions in pottery, decorative smoking pipes, and other artwork.

The Historical Park's headquarters is located at the Mound City unit. Twenty-three mounds dot the sixty-eight-acre enclosure. Most of the mounds are reconstructions, our guide Sue explained. Early Euro-American farmers plowed across some of the mounds, the seasonally repeated action wearing them down over the years. And in 1917, the US government built Camp Sherman at the site as a training ground for soldiers headed off to World War I. While they spared a handful of the mounds, they shaved off most others, with the camp buildings constructed on top of the sheared mounds. The constructions did not intrude into the base of the mounds, however.[6]

The mounds' unique construction made them "rebuildable" after the generations of abuse. Burials at Mound City were not *in* the mounds, but beneath them and thus not destroyed by the disruptions. Each site, Sue explained, began not as a mound but as a funerary building in which bone bundles or cremains were interred within the earthen floor. When the floor space was full or its function completed, the building itself was dismantled and burned, and an earthen mound placed over the site, constructed with alternating layers of clay and sand, then topped with gravel, all of which was designed to shed water and last through the ages. Then a new funerary building was constructed, and the process repeated.

Sue pointed to the rectangular berm that enclosed the space. We guessed that it might have been a defense wall for the community. But no one ever actually lived here, she explained. There was no "city" at Mound City. Hopewell peoples lived in scattered two- or three-hut gatherings in the local vicinity, but came together here for centuries to bury the dead, and to build and maintain the mounds. The perimeter berm appears to have been added when the mound and burial work was, for whatever reason, deemed completed.

Also found in the mounds and among the burials were artworks from materials originating across North America, indicating sophisticated trade patterns. Copper—perhaps from the Upper Peninsula of Michigan—was patterned into intricate forms: a copper headdress, copper hands, copper stars, copper antlers, and copper falcons. Mica from North Carolina mountains was thinly sliced, like onionskin, into the shape of hands and abstract swirls. Ceramics and pottery utilized designs traceable to indigenous peoples from the Appalachian Mountains.

Hopewell art took many forms. Artisans shaped ceramic pipes in the forms of turtles, wildcats, falcons, and squirrels. A cache of these was discovered in the Mound of the Pipes, all ritually broken as if to free the effigies' spirits within the grave.

Dianne and I walked among the mounds after Sue had finished the tour. Taller than the Upper Mississippi mounds, the largest here reached seventeen feet with a ninety-foot circumference. One unreconstructed burial site demonstrated how Mound City looked after the Camp Sherman leveling. We passed through an opening in the perimeter wall and walked to the edge of the Scioto River that connected Mound City to other Hopewell sites and beyond through confluences with the Ohio and Mississippi Rivers.

Archaeologist John P. Hancock writes that, in Hopewell cosmology, "The earth where we live is brought into being between the sky world (of the Sun, the Thunders, wind, and lightning) and the beneath world (of the Moon, water, and the Underwater Panther)."[7]

There was balance here: earth, sky, water, and what lay buried.

---

When we arrived at our Lake Hope State Park campground twenty miles southeast of Hocking Hills State Park, we stopped at the office to sign in for our reserved campsite. Three men and the campground hostess eyed us curiously as we stepped inside wearing COVID masks and bearing northern accents. "Have you all found your campsite yet?" asked the hostess.

"Not yet," Dianne said, before asking her for a campground map.

"Lordy! You don't need a map. Just go down the hill, up the next, take a left, and you'll be there!"

The campsite doubled down on the steep southern Ohio hills. With few level spots available for a tent, we pitched ours on an RV parking stall, excellent for someone like myself whose ideal mattress is a cement slab. Bicycling through the campground was gear-shifting heaven or hell, depending on one's perspective.

Over the next few days we found ourselves as much on the border of Appalachian culture as we were in the Midwest. The tiny rural and small-town churches we drove past announced themselves like an Appalachian menu: Shepherd of the Hills Church; Freedom Memorial Church; Christian Prayer Chapel; Heavenly View Tabernacle; Risen Redeemer Full Gospel Church; Cornerstone Baptist.

There was a little bit of the Bible Belt in the campground's no-alcohol policy too.

A roadside stand sold Trump merchandise a year after he'd lost the election. Four-door pickups dominated the gas stations.

But biases have a way of circling back. About to enter a gas station, I encountered a man stretching the length of his grasp, holding a door open alongside his four-year-old daughter. Misreading the moment, I reached for the other door handle, and the father said in disappointment, "Oh, I was teaching her to hold the door open for others."

The places we take root are the bedrock that define, support, and sometimes limit us.

Eventually, though, there are fissures where sandstone gives way to limestone and to new wonders—and even new subterranean flaws.

South of Hocking Hills State Park, the bedrock harbors veins of human treasure and tragedy in almost equal measure. At the edge of Lake Hope sits the ruin of the Hope Furnace, a nineteenth-century iron smelter once fueled by charcoal processed from local forest wood. A community of three hundred iron workers lived near the furnace from 1854 to 1874, tending the flames twenty-four hours a day. Today, the only remains are the trapezoidal hewn-rock smelter with a few tumbled blocks at its base and a top overgrown by forest vines.

This was the northern reach of the Hanging Rock Iron Region, the nation's leading supplier of iron in the mid-1800s. Nearly seventy smelting furnaces scattered along a narrow swath through six southern Ohio counties produced more than one hundred thousand tons of iron per year before the outbreak of the Civil War.[8]

Iron ore could be accessed at ground level here. Red swirls in the sandstone cliff faces indicated the presence of iron. Miners augured holes into the sandstone outcrops and inserted dynamite, then chipped, chiseled, and pickaxed their way to the iron ore seven to ten feet inside the rock face. Ore blocks carted away to the blast furnaces were usually still encased in sandstone, so the smelting process melted off the iron and turned the sandstone into glass slag that still litters the forest floor near the abandoned furnaces.

The southeast Ohio iron industry contributed to another underground phenomenon: many iron mine owners were abolitionists, and smelting furnace towns became stops on the Underground Railroad. Black Fort Settlement on the edge of today's Wayne National Forest was an early 1800s African American ironworks town comprised of Free Blacks and former slaves.[9]

Iron mining in southeast Ohio was short lived. The smelting process required six acres of trees per day per furnace, stripping the region of its forests within a few decades. With forest fuel dwindling, and with new mines opening in a Great Lakes region that had greater iron density and easier shipping opportunities, the iron industry shifted northwest.

The denuded hillsides were highly erodible and difficult to farm. Much of the abandoned southeast Ohio lands were repurposed as state parks and replanted state forests such as the Richland Furnace State Forest (2500 acres), Lake Hope State Park (3000 acres), the Zaleski State Forest (28,000 acres), and the Wayne National Forest (244,000 federally owned acres).

The first Euro-American settlers had claimed that a squirrel could travel tree-to-tree from the Ohio River to Lake Erie without ever setting foot on the ground, but settlement and the iron industry decimated most of the virgin forest. After nearly a century of second-growth forest, however, the squirrels can have at it again, at least in this corner of the state.

Dianne and I drove into Nelsonville on the morning of a day we'd set aside for bicycling. Coming into town we passed the beehive kiln ruins of the Nelsonville Brick Company. Although most of the kilns have disappeared, two remaining circular kilns still sit alongside the road. The intact kiln wore a domed rock roof like a beanie. The other had mostly collapsed into itself.

Clay mining filled the void in southeastern Ohio as the iron industry waned. The clay deposits originated at the bottom of a large, regional ice-age lake. Glaciers to the northwest had blocked the flow of the Teays River, impounding its waters in a glacial lake for sixty-five hundred years. Fine silt deposits at the lake bottom compacted into clay layers up to one hundred feet deep.[10]

The Adena and Hopewell peoples used area clays in their pottery. Large-scale extraction in the Nelsonville area, however, dates back to the first brick plant established in 1870. Nelsonville bricks were used in street construction as well as in building materials. A clay sewer pipe company followed in 1887. Nearby Haydenville followed the same pattern. With Peter Hayden's pig iron business fading by the 1870s, he shifted his interests to clay, firing street-paving and construction bricks and, later on, clay fasteners for electrical conduits. By the 1940s, however, cement was replacing brick in road and building materials, and the clay industry likewise began to collapse.[11]

Next in line, southeast Ohio's coal industry had humble beginnings. Coal was known to be present in the region for decades before it was commercially mined. Early 1800s settlers in Athens County simply gathered surface chunks from the hillsides and riverbanks for their individual use. Some homes had a basement coal mine entrance. But in 1830, as Columbus businessman John Gill delivered four wagonloads of iron stoves seventy-five miles from Columbus to Athens—his enterprise made possible by area iron—he came upon a blacksmith shop in Nelsonville fired with locally gathered coal. On his return trip, Gill loaded up two of his empty wagons with coal, delivering the first shipment back to the booming new capital town.[12]

But little could be done without an efficient means of getting coal out of the remote and hilly region. The first step in spurring the coal industry was the construction of the Hocking Canal alongside the Hocking River in 1832, connecting Athens to Lancaster and, ultimately, to Columbus and the Ohio and Erie Canal. The demand for coal soared as it replaced wood as fuel on steamboats and railroad engines. Coal extraction quickly evolved from household scavenging into a mining industry in the Hocking Valley.

Railroads soon replaced canals for transporting coal. The Columbus & Hocking Railroad came to Nelsonville in 1869 and reached Athens a year later. Within a year, the railroad was moving two hundred thousand tons of coal. Nelsonville, Athens, and a host of other coal towns boomed, with seven mines operating in Nelsonville alone.[13] Sunday Creek, not far from our campground, gave its name to the second largest coal-mining company worldwide. The Sunday Creek Coal Company operated over thirty mines in the Hocking Valley and nearly as many in West Virginia.[14] The nearby Monday Creek watershed was home to numerous mining sites as well.

For decades, coal meant wealth—at least for the mine owners. Nelsonville was the center of the regional coal-mining industry through the 1950s. Company towns like Murray City, Gloucester, Eclipse, and New Straitsville abounded, with their tell-tale look of semi-identical homes on small lots all in a line.[15]

But controversy and tragedy were hauled up alongside the wealth of coal. In 1884, striking workers set fires in numerous coalmines that continued burning for decades, venting from cracks in the surface ground.[16]

Racial tensions exploded in 1874 and 1888 when owners hired out-of-state and nearby African Americans to replace striking Euro-American workers.[17]

In 1872, a mine fire in northern Ohio killed nine men and a nine-year-old boy. After the fire broke out in one part of the mine, the owner sent the boy into its other sections to warn the workers, but all of them perished when the fire consumed the one and only exit.[18] The largest mining disaster in Ohio history occurred in the Hocking Valley in 1930 near Sunday Creek when an explosion killed eighty-two workers as company officials inspected newly installed safety devices. Mining disasters killed another 180 Ohio workers before the end of World War II.[19]

At the start of the twentieth century, fifty thousand coal workers were employed in Athens, Hocking, and Perry counties.[20] But after World War II, mine shaft operations began shutting down, replaced for a while by surface strip mining and mechanized underground long-wall mining, both of which employed vastly fewer workers. Eventually, these mines began closing too. Today, only a small number of coal-related jobs remain in the area.

Meanwhile, over three hundred miles of southeastern Ohio streams bleed orange, white, and green from abandoned mine drainage.[21]

Natalie Kruse Daniels, PhD, professor and director of the Environmental Studies Program at Ohio University in Athens, explains this ecological legacy by first describing the formation of coal in the region.[22] The coal here was formed through the burial and compression of organic matter in an oxygen-deprived environment at the bottom of an ancient sea that was particularly high in sulfates. As a result, Ohio coal is high in iron sulfide (known as pyrite, or "fool's gold"). When mining residue is left behind and exposed to air and water, whether inside the mine or in tailings, says Kruse Daniels, "it weathers, oxidizes, forms sulfuric acid, and releases a significant amount of iron, aluminum, and magnesium."

Mine tailings act like "coffee in a filter," creating significant amounts of Acid Mine Drainage (AMD) as water percolates down through the crushed material, resulting in "environments where things can't live," Kruse Daniels adds. Underground mines can collapse, giving surface waters entrance to the mines where residue remains. The contaminated waters then leak back into the environment from the mines. In a chapter titled "The Legacy of Appalachian Ohio Coal Mining" in *Surviving to Thriving in Appalachia*, Kruse Daniels and her coauthors write, "Creeks and streams flowing with AMD provoke local schoolchildren to reach for orange crayons when drawing their environments."[23] She notes that polluted streams often take on the color of "tomato soup."[24]

Mitigation efforts are helping to restore some streams. Restoration efforts led by the Ohio Department of Natural Resources with help from Ohio University in Athens have involved spending over $30 million since 2000 to decrease the acidity in 193 miles of streams to the point where fish can again live there. At least ninety-three restored stream miles once again meet state standards for "healthy streams."[25]

Although the region has had a high poverty level since the collapse of coal mining, there are vestiges of economic recovery as well, especially in communities that lean into outdoor recreational opportunities in the stunning hills of southeast Ohio. One example is the Baileys Trail System that, when completed, will offer eighty-eight miles of world-class mountain biking through former coal mine country near Chauncey, Ohio, just north of Athens, and through the Wayne National Forest. Thirty-one miles are already open, some with mining-themed trail names such as the Coal Train, the Salt Works, and the Gob Pile. In 2021, the state of Ohio invested $2 million—and local governments kicked in funds as well—for the trail system that will be the largest east of the Mississippi. Some proponents project the trails will bring in $40 million of visitor spending over a ten-year period.[26]

The same early evening that we visited the Hope Furnace, we rode out the rest of the daylight with a leisurely drive. We crossed Monday Creek and we crossed Sunday Creek. We stopped at the Hope one-room schoolhouse that once served the Hope Furnace community. We looked for the pioneer cemetery, but couldn't find the way in through the forest. We drove the winding Wheelabout Road up from the valley to the high ridge, dodging potholes and bouncing along ruts.

The Hope-Moonville Road took us to the Moonville Tunnel in the Zaleski State Forest. We turned into a muddy parking lot. One nearly broken-down car sat amid the puddles near the access to the sixteen-mile-long Moonville Rail Trail. The trail led to the railroad tunnel that cut into the hills.

Moonville was home to one hundred villagers in the late 1800s, coal miners mostly. Workers walked the tracks to and from the local mines. The town dried up as the coal mines closed, with the last family departing in 1947.

The trail crosses a rusting iron trestle bridge over muddy Raccoon Creek before arriving at the tunnel. The Moonville Tunnel is still part of local lore as the haunted site of six accumulated deaths until the railroad ceased

Southeast Ohio was a major coal-mining region until the mid-twentieth century. The Moonville Tunnel marks the location of the long-disappeared coal-mining community of Moonville. The Moonville Rail Trail now runs along the former railbed that workers once walked to get to the mines.

operating in 1988. The hundred-foot tunnel sports a handsomely bricked edifice announcing in large letters, "MOONVILLE." Above, below, and beside the town name, and inside the tunnel as well, layers of timeworn graffiti and the encroaching forest veil Moonville's story.

Advancing darkness turned us away, back to the battered road, and back to the Lake Hope campground.

It was eerie, all right. There was something haunting at the Moonville Tunnel, but it wasn't about railroad deaths. It was something from the underground itself. It had to do with the coal mines.

It came as no surprise to us that a bike trail should pull it all together. Before we left southern Ohio, we spent the better part of a day on the Hockhocking Adena Bikeway that runs twenty-two miles from Nelsonville to Athens on the former bed of the Columbus-Athens railroad. On the way to the trailhead at Nelsonville, we passed the dome-shaped brick kilns on the edge of town. The trailhead sits astride the last remnant of the railroad, where the Hocking Valley Scenic Railway now runs themed tourist train rides through the countryside, pulled by vintage engines. The paved trail largely runs along the serpentine Hocking River. It passes the Eclipse Company Store, a coal-company era supply store now turned fashionable craft beer hall, where we returned later for an afternoon drink. We passed the river plains where the Adena culture built mounds from 800 BC to AD 1. We cycled into Athens, bedazzled by the scenic route along the river and by the university town's architectural salute to the days of coal wealth. And then we retraced our route through it all as if digging deeper into the landscape.

Ancient, indigenous peoples strove to keep the upper, lower, and middle worlds in balance. Archaeologist Clark Mallam once described Native American mounds as "metaphorical expressions that stress the idealized state between nature and culture—harmony and balance."[27]

The Hocking Hills' geological wonders keep the balance as well with underworld caves and rock shelters providing refuge to the middle world's living peoples. Their cliffs cut across the upper world skies. Their prancing waterfalls bind the sky and earth, and occasionally dip below.

But what happens when the balance is upset? When extractive industries deplete buried ore and strip the forest? When boom leads to bust, and people are without jobs? When the imbalance poisons streams?

And can second-growth forests and the visitors they attract ever restore the balance between the needs of the natural world and people who must work for a living?

Maybe these are the questions that haunt the Moonville Tunnel.

# 4

# Indiana

## *Hoosier National Forest: The Crossroads*

In the quiet of an Indiana woods, after Dianne and I have finished talking, we notice the slight gully intersecting the hiking trail and continuing on the other side. Wide enough for a compact car, it is overgrown with scrub trees, but its U-shaped track through the woods is still faintly visible.

We step aside and listen, not for anything present, but for what's long gone: hooves pounding the soil and raising a dust that strings along to the horizon. Snorts and grunts and bleats bellowing past.

We find ourselves at a crossroads of the past and present, hiking alongside the Buffalo Trace in the Hoosier National Forest in southern Indiana. When bison still roamed freely, they rushed along this path after crossing the Ohio River at a shallow ford each spring and dispersed throughout western Indiana and Illinois. In the fall they retraced the path back to warmer Kentucky grasslands. Their weight and numbers pummeled and pressed the pathway into the gullied depression still visible in parts of the woods.

The Hoosier National Forest covers 203,000 acres of south-central Indiana. In a state known for flat, sprawling cornfields and long straight roads, the forest landscape rises and dips from the Ohio River valley to just south of Bloomington.

The forest, largely a second-growth replanting, was created out of the Depression-era heartbreak of failed farming communities. At a crossroads of broken dreams, a new course was charted.

Indiana calls itself "the crossroads state." In the 1800s, horse and wagon pathways intersected in the state as Americans and immigrants plied west and south. In modern times, the city of Indianapolis claims the title all for itself based on the junctions of Interstates 65, 69, 70, and 74. Dianne and I know this. We hit traffic from all four routes as we pushed through the city.

Our plan was to cross Indiana en route to Ohio and return ten days later to the Hoosier state. But we stopped first in West Lafayette to meet up with Laura, a former student of mine, and her fiancé, Tim (entering their own crossroads of life). A mature and thoughtful couple, they listened to stories from our then-thirty-eight-year marriage.

West Lafayette sits in Indiana's Central Till Plain, a Corn Belt region with thick, black soils that derive from glacial till. The steep banks of the Wabash River on the city's east side were chiseled by those same glaciers' meltwaters.

Human history intersects the glacial story in the Wabash River valley. The first peoples inhabited the valley as the ice retreated. In the counties directly northeast of West Lafayette, archaeologists have uncovered twelve Paleo-Indian Period (12,000–10,000 BC) sites in the river valley and twenty-four within one mile of the Wabash. Other sites date back to Archaic Period (10,000–1000 BC) and Woodland Period (1000 BC–AD 1200) Native Americans.[1]

We felt the heaviest convergence of history, though, as we walked the grounds of the Tippecanoe Battlefield Park where the dream of a united confederacy of first nations came to an end in 1811. In the years leading up to the battle, Shawnee chief Tecumseh had negotiated a loose association among Potawatomi, Kickapoo, Winnebago, Sauk, Ottawa, Shawnee, and other tribes to collaboratively stem the flow of American immigrants into midwestern lands. The heart of the confederacy lay at Prophetstown, just north of today's West Lafayette. But the dream of united resistance and peaceful coexistence crashed at the Battle of Tippecanoe when nervous American troops took advantage of Tecumseh's temporary absence and planned an attack against Prophetstown. Sensing trouble from the amassed American troops, though, the tribes struck first, but lost the resulting battle, setting in motion the end of the Indian confederacy.

At the commemorative site rest several soldiers' gravestones ("Capt Wm C. Baen, Killed Nov 7, 1811") and a plaque that reads "Loss: Americans Killed, 37; Wounded, 151; Indian Loss Unknown."

Unknown.

The Potawatomi Trail of Tears crossed through the battle site decades later, during their forced removal from Indiana to Kansas in 1838.

A tone-deaf sign at the park redirects visitors' attention to the natural beauty of the park, noting the three-hundred-year-old white oaks and heralding "a carpet of spring beauties, patches of fuzzy pussy toes, and little colonies of trilliums."

Perhaps, though, that is the pathway to healing.

After lunch with Laura and Tim, we drove southeast across Indiana on the way to Ohio, and several days later returned to Indiana by way of the Southern Hills and Lowland Region that marks the boundary of both recent and older glaciation. Here lie the unglaciated hills of the southern third of the state. The region is rich in deep layers of limestone and underground streams and caves. This was a region of old-growth oak and hickory forests up until 1800, whereupon it was quickly cleared for farming, and almost as quickly saw its thin soils washed away.[2]

Before arriving at the Hoosier National Forest, however, we first stopped at the fifty-thousand-acre Big Oaks National Wildlife Refuge. Here, too, was a crossroads, this one of war and peace.

The Big Oaks Refuge occupies the former Jefferson Proving Ground, established in 1941. The army base displaced twenty-five hundred residents on a few months' notice on the eve of World War II to establish a testing ground for artillery, landmines, and other weapons.[3] Like many such army bases, the Jefferson Proving Ground eventually outlived its functionality and closed in 1995, except for an eleven-hundred-acre section still retained for Air National Guard training.

The remainder of the proving ground began a new life as a wildlife refuge with wetland, forest, and grassland habitats, and is a designated Globally Important Bird Area.

But there is always something lurking in the woods. Because its test-firing past has left the site with buried, undetected explosives, entry to the refuge is guarded, and hikers, fishermen, and site-seers must first view a film warning them of what may lie on the surface or just below. A sign warns that "of approximately 24 million rounds fired on Jefferson Proving

Ground from 1941 through 1995, approximately 1.5 million could be Unexploded Ordinance (UXO)." It continues: "Be aware that UXO can be encountered at any location on the refuge. DO NOT TOUCH ANY METAL OBJECT."

We were planning to explore Indiana's underground wonders, but this wasn't quite what we'd had in mind.

---

Bedford, Indiana, sits at the edge of the Hoosier National Forest, and is the location of its administrative headquarters, although a secondary headquarters in Tell City services the southern half. We set out to talk with Visitor Services Information Assistant Mia Gilbert, whom we met at a wooden picnic table under the pines that tower over the Brooks Cabin in the Hoosier National Forest.

The setting was ideal. The national forest includes several subunits, one of which is the Charles C. Deam Wilderness Area. The Brooks Cabin sat at the end of a long and dusty gravel road within the Deam Area. It was a cool morning, the last such blush we'd have before the onset of a hot summer.

Mia met us at the turnaround circle with an enthusiastic wave and smile. She'd been with the Hoosier National Forest for about two years, after having studied anthropology as an undergrad in West Virginia and working at the Tongass National Forest in Alaska. She was currently pursuing a master's degree in public affairs and social media. Her job at the Hoosier National Forest involved communicating with the public, keeping them informed about any major plans (such as upcoming prescribed prairie or forest burns), and getting feedback "so we shape our plans just right." [4]

Mia described the Hoosier National Forest's historical beginnings, a story we dug into more deeply later. By the early 1770s, British American settlers from Kentucky began moving into southern Indiana, even though the Proclamation of 1763 prohibited their settlement west of the Appalachians. The settlement pace quickened when the Land Grant Act of 1800 offered property for $1.25 per acre. Indian treaties—made, broken, and reimposed—between 1783 and 1840 further intensified Euro-American settlement.[5] The old-growth forest fell quickly in favor of the plow.

But the forest soil was thin and the land was both hilly and rocky. Erosion control was not yet widely practiced, and as the soils washed away, area farms lost productivity and economic viability. Add in the droughts of the 1930s and the overall effects of the Great Depression, and all too soon landowners were abandoning their farms. Congress granted the National Forest Service the right to buy up delinquent lands from owners willing to sell. By August 1935, the Forest Service was inundated with two thousand

The Brooks Cabin sits at the edge of the Charles C. Deam Wilderness Area within southern Indiana's Hoosier National Forest.

such offers, totaling two hundred thousand acres, more than it was able to purchase at that time.[6]

Mia noted that the Forest Service immediately began reestablishing the forest to control erosion. Needing to quickly reestablish a forest, they mostly planted eastern white pines. "We are slowly trying to get it back to native hardwoods," Mia said, as the now-mature pines are being harvested and replaced.

Dianne often accompanies me to these interviews. We make a good team. I take notes the old-fashioned way—by handwriting—and that means I easily fall behind the conversation. Dianne fills in the spaces with questions and comments of her own while I catch up. She was curious, she said, about the thirteen-thousand-acre Deam Wilderness Area. What is a wilderness area within a wilderness?

Mia explained that while a national forest is, by definition, less intensely altered for the general public than is a national park, a wilderness area is meant to be left even more hands-off. Even most forest management practices are barred there, such as felling nonnative pines and replanting native hardwoods. "Our mission here is to leave it exactly as it is. We humans are not meant to stay," she said.

Wheeled vehicles are not allowed in the Deam Wilderness Area—no cars, trucks, or bicycles—except on a few remaining roads. Trail maintenance, for example, is accomplished using Ruthie and her teammates—Forest Service mules (the four-legged kind)—pulling carts loaded with tools and materials. Such restrictions help preserve the solitude of a wilderness area, Mia explained, although some occasions call for creative work-arounds. When one trail suffered a particularly destructive washout, Forest personnel sought and received permission to drop a load of gravel by helicopter from above the tree line, thereby not technically treading on Deam Wilderness Area soil.

Mia stayed with us for over an hour and gave us pointers for viewing the Deam Wilderness. First, of course, we should climb the Hickory Ridge Lookout Tower, a 150-foot, 1936 CCC steel construction that, along with seven other such towers, was at one time individually manned to watch for forest fires. The area was still transitioning from farm country in the 1930s, so at the time of construction the tower also overlooked eighty farms and homesteads, all of which have now been subsumed into the forest.

Departing from our conversation with Mia, we drove directly to the tower. We climbed the tower's 123 metal stair-steps, with Dianne bounding confidently ahead of me as I contemplated each step, both up and back down. At the top, a graffiti-laden, seven-foot-square roofed platform offered some protection from the sun above the tree line. No longer serviced since the 1970s, the Hickory Ridge tower is the last of the eight still standing.

As we drove back down Tower Ridge Road, I noticed from the corner of my eye a hidden sign announcing the Town of Todd Cemetery. I'm attracted to old cemeteries, so we pulled in. While there were a few modern burials, still decorated, much of the cemetery hearkened back to the farming community that predated the national forest. Small, weather-beaten headstones poked from the ground, leaning in every direction. Many were unreadable, but a few still told their stories. There was Robert Crough, the toddler who lived from 1927 to 1930. The headstone of Louisa E. (illegible last name) is hard to read, but one interpretation of her dates puts her in her twenties, a likely death while giving childbirth. Private William Hicks, on the other hand, had long outlived his battle days, finally succumbing at fifty-three years old, his life spanning from 1829 to 1882.

This was a crossroads of life and death. Death came frequently, back then, to children and women in labor, and to soldiers in interminable wars. But if a Private Hicks managed to survive war alongside his North Carolina infantry comrades, he might move to Indiana and grow reasonably old.

But the cemetery was also a crossroads between the farming past and the forested present, where the only sound today was the wind licking through the trees. A cemetery in the middle of the forest? This was new to us. But soon enough we began seeing them all around as we drove or hiked, and we spotted even more on the maps: the Callahan Cemetery, Elkinsville Cemetery, Fleetwood Cemetery, Lucas Cemetery, Tinscher Cemetery, Seton Cemetery, Luxenburger Cemetery, Colby Cemetery, and Talley Cemetery. Drop a pin on the map: there's one. Drive slowly through the forest: there's another.

But in the end, not such a bad place to rest.

Pockets of privately owned land and houses still dot the Hoosier National Forest. Roads vary from the well paved and well traveled to sketchy backroad gravel lanes. They twist and wind through the hills, as if the builders had changed their minds and directions during construction. Back-country lanes sometimes lack signage, don't appear on the printout map, and are terra incognita on Google Maps.

Dianne prides herself on being the navigator. I just drive. She knows better than to tell me to turn east or west, as that could be right or left or behind me. So when we get lost, she takes it as a personal affront. I just turn the car around.

"Do you remember this road from this morning?" she asks after we retrace our route and turn left onto an unmarked lane.

I shrug. "I don't remember it from five minutes ago."

But we intentionally spend part of our day wandering and exploring in between planned hikes. One such wandering takes us past an active quarry that spreads a coat of dust across our windshield, car, and the bikes on the rear rack.

Lime dust. We are in the land of Indiana limestone, especially from Bloomington to the northern reaches of the forest. Known by geologists as Salem limestone, the sedimentary rock was laid down in shallow equatorial seas 340 million years ago. The shallow depth was ideal for small, shelled sea creatures, whose casings accumulated and packed tightly into the sediment on the ocean floor and, over time, transformed into limestone.

While limestone is one of the world's most common bedrocks, especially throughout the Midwest, Indiana limestone has qualities that make it particularly treasured as a building material. It is soft and easy to quarry, but once exposed and dried, its surface becomes "case hardened" and less subject to weathering than other limestones. With no "preferential direction of splitting," it can be shaped, sawed, and carved easily as well.[7]

As a result, Indiana limestone has often been the stone of preference for the nation's most prestigious buildings. Architectural works making use of Indiana limestone include the Empire State Building, the Pentagon,

Southern Indiana's limestone bedrock harbors underground treasures such as the Marengo Cave.

Yankee Stadium, Washington Cathedral, the Chicago Tribune Tower, the Lincoln Memorial, and twenty-seven state capitol buildings. And while the industry's heyday dates back to the 1920s, several active quarries remain, thanks in part to the thickness of the Salem limestone seam, which can run from twenty-five to sixty feet deep.[8]

Thick limestone beds, especially when they occur just beneath the land surface, also create a geological phenomenon known as karst landscapes. Over time, limestone is soluble. Rainwater and snowmelt will percolate

down from the soil till they reach bedrock. Then they follow along vertical and horizontal seams in the limestone, gradually enlarging them by dissolving the rock, and creating underground streams and caves. A limestone cap over the stream or cave may collapse, creating a sinkhole at the ground surface, giving water, humans, and livestock direct downward access to the cavern beneath.

Such, too, is the landscape of my home in northeast Iowa, in the Driftless region, where limestone sits directly under the soil. I know any number of places where, in the thick of winter, I can feel the warm breath of cave air rising through a hole in the ground.

That affinity with our home landscape led us to the Wesley Chapel Gulf, a 187-acre tract within the forest where at least as much is happening below ground as above. Here the Lost River runs for twenty-three of its eighty-five miles underground, resurfacing occasionally where the limestone cap above it has collapsed or where underground pressures force it upward.[9] At Wesley Chapel, a series of sinkholes have eroded into each other, leaving a hollow where the Lost River bubbles up, and into which it disappears underground again downstream.

The pool through which it flows varies in depth and clarity according to season. Fed by snowmelt in the early spring, it runs deepest and clearest at this time of year. We are here instead in summer, beating back the weeds and watching for ticks on this little-used path, and when we finally come upon the pool, it is not at its most pleasant state. It is green and murky, and hardly inviting. But we poke around a while until we find the little gash at the pool bottom from where the stream arises. The pool's exit is overgrown, so we can't see how far the stream goes before it dives back underground.

In the 1800s, local churches brought new congregants down to the Wesley Chapel Gulf for baptism in these waters that welled up from the underworld.[10]

Since karst also means caves, we drive a little east of the Hoosier National Forest to the Marengo Cave. I can be a little stodgy when it comes to commercialized caves, and Marengo is no exception with its grounds also hosting "pedal karting," gemstone mining for kids, and a "mega maze." But then again, my kids are grown, and I've got no one to keep entertained except my inner nerd. Sugarcoat it all you want with fun asides and silly names for cave features, in the end there is still the geological and mystical experience of a cave. And I am thankful to the owners for making it accessible to the likes of myself, the kind of person not likely to don spelunkers' gear.

Marengo Cave, a National Natural Landmark, was discovered in 1883 by two young farm siblings, fifteen-year-old Blanche and eleven-year-old Orris Hiestand, who crawled into an opening at the bottom of a sinkhole with candles and string. They and later explorers discovered almost five miles of winding caverns. In addition to the usual stalactites (hanging *tight* from the ceiling) and stalagmites (growing up from the floor), other geological formations developed there. These formations include soda straws (short, thin, tubular calcite growths suspended from the ceiling); flowstone (from seepage weeping down a cave wall); draperies (thin, translucent sheets hanging down from a ceiling); cave popcorn (small, knobby clusters of calcite about the size of grapes); helictites (twisted collections of angular calcite that look like clumps of mineralized worms—think "hell"); and rimstone dams (calcite buildups rimming the edges of interior ponds). In numerous locations, shallow reflecting pools give the illusion of unlimited depth, as their still waters perfectly mirror the formations above them.

Cave formations call out to be named and given stories. To wit, in a wall of flowstone I saw the shape of an arm and hand reaching down, as if poking around on the cave floor for something dropped and lost. Gollum searching for his lost ring.

Life-forms within the cave offer their own little shop of horrors. Blind, translucent isopods scuttle along on shrimplike legs searching for microscopic food in the cave's streams, hiding out from equally blind, eyeless, pigment-less fish on the hunt. Millipedes and beetles explore the mud banks. Tiny, white-gray springtails may jump out at you when disturbed. Troglobitic spiders guard the natural entrances to the cave, catching and caching luckless insects that fumble into the openings.

At the far reaches of the cave, a hard sandstone cap overlies the limestone, and here the calcite formations end abruptly as the sandstone prevents surface water from entering the cave. This part of the cave is drier, with clay floors. The underground cavern here has hosted ministers sermonizing from a loft called the Pulpit (not quite heavenly), musical performances (oh the echo!), and the occasional underground wedding (insert your own comment about the couple from hell).

The walls here, too, are plainer, and have over time attracted some abuse. Graffiti is illegal in Indiana caves, but historical graffiti seems inexplicably less offensive, almost romantic. Just why did the Retail Grocers Association of New Albany, Indiana, think it a good idea to scrawl an advertisement for their wares in the darkness of a Marengo cave wall on September 21, 1913?

Rain and meltwater enter the Marengo Cave through sinkholes or seeps down through cracks in the limestone. Dripping interminably from the ceiling, cave water eventually forms an underground stream that emerges on the landscape five miles later at a spring on Whisky Run Creek that flows into the Blue River that will make its own crossroads with the Ohio.

Tell City lies just beyond the southern edge of the Hoosier National Forest along the Ohio River and serves as administrative headquarters for the forest's southern unit. There we met Visitor Services Information Assistant Alexander Johnson, who graciously welcomed us into the conference room. [11] He would have preferred to have met us in the forest itself—as Mia did—but on this day the office was short staffed and he needed to be in the building. But it was good to meet here, too, where forest maps, public communications, and habitat restoration plans adorned the walls.

Alexander pointed out a key difference between the Hoosier National Forest and those of the West. "Out west," he said, "forests were land that other people didn't want. Here, people already owned these lands, but drought, bankruptcies, and foreclosures drove them away." Erosion here had been so pronounced that it was impossible to make a living growing crops.

In the early years, the Forest Service "had planted tons of nonnative eastern white pine" in the newly minted Hoosier National Forest. "It served its purpose, it held the soil," he said, "but now it's time to reestablish the hardwood forest." Not far from our campsite at Indian Lake, he said, a recent harvest of matured eastern pines had created the opportunity to replant with native hardwoods, which in turn will provide more habitat for native animal species.

Timber sales are a reality in national forests. Alexander referenced the 1905 Mission Statement written by the first Forest Service director, Gifford Pinchot: "Where conflicting interests must be reconciled, the question shall always be answered from the standpoint of the greatest good of the greatest number in the long run."[12] And sometimes the greater good involves harvesting trees. Timber sales help fund the agency and services within the forest but can also serve environmental purposes, such as reducing overgrowth, creating fire breaks in the forest, and creating open canopy necessary for new growth, especially for hardwoods. "We are a conservation, not a preservation, agency," Alexander added. At a comparatively small national forest like Hoosier, though, logging is not widespread.

Alexander told us more: about prescribed woodland burns to keep the understory in check; about the Pioneer Mothers Memorial Forest and the Hemlock Cliffs we ought to visit; about Buzzard Ridge above the Ohio River; and about the Barrens, a section of sparse vegetation in the middle of the forest. He told us about the famed local bourbon, Buffalo Trace, too.

Alexander trained as an undergraduate at the University of Illinois-Carbondale to become a history teacher. But he also grew up loving nature and camping, so he does his teaching these days with the Forest Service. "I still get to visit schools, but this way they get excited when I show up," he laughed. He meets school groups in the forest as well. "Even if they don't specifically remember what I'm talking about, they develop a passion for the outdoors."

We'd learned much in our hour with Alexander. Armed with several additional recommended sites to visit, we took our leave.

Among our intended visits was a hike at Lick Creek, a long-disappeared settlement founded by freed Black families in an east-central section of today's Hoosier National Forest. First, though, after leaving the Tell City Forest Service office, we drove down to the Ohio River shoreline at the southern edge of town. Here was a crossroads I hadn't confronted before in the Midwest, the pre-Civil War boundary between Free and Slave states.

It's not as if the Mississippi River that runs alongside my home in northern Iowa was free from historical strife. The Mississippi marked the eastern boundary of where the Sauk and Meskwaki had been pushed by 1830. When tensions erupted in the 1832 Black Hawk War, the Mississippi soon ran red with blood when the Sauk were massacred trying to swim to safety.[13] Soon after the war and new punitive treaties, lead miners crossed the Mississippi near my home, laying claim to lands now opened to "settlers." They noted the sophisticated layout of the abandoned Meskwaki village and then burned it.[14]

So my own Mississippi had carried stained historical waters. But I'd not confronted the historical Free state / Slave state boundary thus far in my Midwest experience. Crossing inside the closeable floodgate wall, we parked and walked along the Ohio River. Here the Ohio was almost but not quite as wide as the Mississippi at my home. I imagined standing on the opposite shore in the nineteenth century, in Slave state Kentucky, contemplating the Indiana shoreline and wondering how close or how far away it must have looked. There was no floodwall then, of course, but surely the gate to freedom must have seemed, in equal measures, both open and shut.

A long road took us back toward the former Lick Creek settlement. Although many forest roads wind up and down and circle around through the thick woods, County Road 250 is a straight, short arrow off the paved highway, lined with sun-bathed farm fields and handsome houses on either side. As we neared the end of the lane and found a few slots of off-road parking, a signboard pointed to the historical settlement of Lick Creek that once lay farther down the hiking path in the same direction from where the road had ended.

Freed Black families started coming to Lick Creek from North Carolina around 1811, since even the possession of "Freedom Papers" didn't guarantee safety in slave-owning states. Elias and Nancy Roberts arrived in 1823 with a document proclaiming they were "free and entitled to all the rights and privileges of white persons." Invited and welcomed by the still-active Quaker Church of Lick Creek, freed Black families began purchasing land from the government. By 1850, the community had expanded to 260 Black settlers in ownership of twenty-three hundred acres.[15]

By 1853, however, the state of Indiana required Black adults to register by name, place of birth, and physical description. Elias Roberts, thirty years a citizen at Lick Creek, was the first person whose name was so recorded. Prospects grew more ominous as the Civil War broke out across the nearby Ohio River. In 1862, with the war intensifying, seven Black families sold their land and moved farther north. Their departure was abrupt, evidenced by Solomon Newby, who sold his land for half the price he'd paid for it sixteen years earlier. Six of the seven families fled to Canada. Before long, Lick Creek was abandoned and the forest grew back.[16]

Dianne and I walked a long time along the trail that wound through the remains of the community. We tried to imagine farmhouses and barns and fences and the community church, but our untrained eyes couldn't discern them in the underbrush. The land sloped quickly from the hiking trail, probably not the choicest farmland in the county. The path to the cemetery was overgrown, and so we left it to its own.

I am a connoisseur of quiet, but this was a quiet that spoke too much.

Another section of forest that likewise spoke its history quietly was the Pioneer Mothers Memorial Forest. This eighty-eight-acre tract was first purchased in 1816 by Mary and Joseph Cox, who'd come to Indiana from Tennessee. They'd settled nearby on land they intended to farm, but bought this acreage to preserve as forest. The tract remained in the family for 124 years until the 1940s, when later descendants sold it to a Louisville lumber company for logging. But a community effort arose to repurchase the land to honor the Cox's original vision. A large donation came from the Indiana Pioneer Mother's Club. When the tract was remitted to the Forest Service, it came with two stipulations: that it be named to honor pioneer women, and that no trees ever be cut from the property. Today it is one of the few old-growth timber stands in the state of Indiana.

North of the Pioneer Mothers trail sat a small parking lot and a blocked-off road. The roadblock offers a paved, flat, wheelchair accessible, quarter-mile route through the woods, across an old stone bridge, and through an adjacent forest that is similarly protected. At the end of this lane the hiking trail rises into the forest where, soon enough, an eight-foot-tall by fifty-foot-long limestone wall adorns the woods, engraved with a large-lettered etching announcing the "Indiana Pioneer Mothers Memorial." The wall was built in 1951, long enough ago for the forest to begin reclaiming the area. The wall overgrown by woods seems on the one hand forgotten and neglected—as many pioneer women had been—but on the other hand seems appropriately immemorial and eternal, suggesting that these woods existed long before the written or oral word and will live on equally long into the future.

Maybe the long-gone pioneer women would best understand the serenity of accepting that one's own monument will fade into the woods. And, for that matter, so might the Native American women who preceded them and have no monument at all.

Time may stretch endlessly, but the Midwest likewise stretches too far to be contained within a single time zone. Indiana is a crossroads in this regard as well. Drivers crossing the northwestern counties from west to east will find that central time changes to eastern time three counties in. Most of the state lies in eastern time, except for another swatch of six counties in the southwest corner. These two central time anomalies allow for better alignment with Chicago to the northwest, and in the southwest ensures that the regional hub of Evansville, IN, shares the same time zone as its outlying neighbors in western Kentucky.

I mention this mostly because the time zone line doesn't move north-south through the Hoosier National Forest, as one might expect, but cuts across the forest in an east-west path. Heading north into the forest, one crosses into eastern time. Heading south—as we did each evening while returning to our campsite—meant crossing back into central time.

Were we more attuned simply to sun time, it wouldn't have mattered a whit.

But there were traces of southern time, too, if such a thing exists. One evening—tired of camp-cooked meals—we headed north into eastern time to find a restaurant where the waitress treated us with a southern familiarity, referring to us frequently as "Hon'" and "Dear."

Then back south into central time to our campsite, a beautiful location amid the eastern pines. Every direction, it seemed, all at once.

In the southern half of the Hoosier National Forest, limestone bedrock gives way to sandstone with a personality all its own. The Hemlock Cliffs, one of the best examples of a beautiful sandstone shelf, is a watery canyon that Alexander had guaranteed would be five to ten degrees Fahrenheit cooler than the prevailing forest. It is one of the Hoosier National Forest's most-visited sites. Known as a box canyon, it drops away from the surrounding landscape at a U-shaped overhang and deepens alongside steep and narrow sandstone walls until it finally opens at the far end into a sloping valley.

From the parking lot, the trail quickly disappears into a crevice in the bedrock and twists and sinks into the two-hundred-foot canyon. The sandstone walls of the descent are pocked and honeycombed from weathering, like a hard lace.

Dianne was moving more quickly and with more agility than me this morning. The descent was a bit challenging, and I was glad to have brought along a hiking stick. I was shooting photos, too, slowing me even more. I took photos of the canyon. I took photos of the sandstone bedrock. I took several photos of Dianne, too, as she moved ahead of me and finally sat on the bottom stair of the trail to wait. And then I simply paused to look at her.

After I caught up, we wedged ourselves between some boulders to catch a view of the U-shaped cliff overhang. A harder layer of sandstone had prevailed against the softer layers underneath, which had eroded back. A small stream plummeted over the ledge, scattering the morning sunlight within the droplets as they fell. The droplets landed with an ongoing patter. We scrabbled into the arc-shaped rock shelter behind the waterfall and looked again down-canyon. The dark edges of the shelter's ceiling framed our view.

And that's when I thought back to our descent into the canyon just minutes before. We usually think of crossroads as points where a decision must be made between two paths. And I suppose that's what it really does mean. But I like to think, too, that there are moments in life where something special is happening, and you either recognize it or you don't. I think that's another kind of crossroads. And there I had been, thirty-eight years into marriage, simply gazing at my wife as she sat waiting, looking down the canyon lost in thought.

---

We were walking hand in hand, I believe, when we noticed that the Buffalo Trace had crossed our path. Then we stepped down into the gully on either side to check it out.

The bison had laid down a trail that others followed. Native Americans used it as a traveling path. The French used it in the 1700s to access their trading post at nearby Vincennes. British and then American army units marched along it. Settlers disembarking from boats on the Ohio River—at the ford where the buffalo had crossed—traveled the route in wagons, heading west, adding wheel ruts to the hoofprints. Stagecoaches soon followed, with inns and taverns popping up alongside the trace to cater to their passengers. The US mail moved along the route, trimming the delivery time from the east coast from twenty days to six. US Highway 150 was laid down parallel to parts of the trace.[17]

The last bison spotted on Buffalo Trace was in 1799.[18]

Most of the Buffalo Trace is gone. But here in the woods it runs beside us, crosses our path, and disappears again into the forest. I imagine the long line of dust that the bison must have kicked up with the weight of their hooves, a string that would have lingered long after their passing and tied together the horizons and everything beneath the dome of sky.

Crossroads take many forms. I'm not so interested in the interstate variety, where mindlessly speeding autos buzz past each other like a swarm of gnats. I'm more interested in the crossroads where real life is lived. Here in the Hoosier National Forest, real people tried to farm and came to the brutal conclusion that the soil was against them. Some were driven away by prejudice. Others took a new path amid the cemeteries of lives lived and dreams lost, and they planted a national forest.

In the end, a crossroads is not so much about the intersections where we make fateful decisions, but about recognizing the rest of the way forward, the path that is lined with ancient, solid bedrock, and with hope and love.

# 5

# Kansas

## *The Flint Hills: Red Buffalo*

"Red Buffalo is what the Plains tribes called prairie fire," said Heather Brown, chief of interpretation and visitor services at the Tallgrass Prairie National Preserve in the Flint Hills of east-central Kansas.[1] The name depicts the speed and fury of fire racing across the Plains but also signifies life returning to the dormant prairie each spring.

Author William Least Heat-Moon put it another way: "The four horsemen of the prairie are tornado, locust, drought, and fire, and the greatest of these is fire, a rider with two faces because for everything taken it makes a return in equal measure."[2]

The only thing sweeping across the prairie as Dianne and I headed southwest out of Kansas City on Interstate 35 on a hot July afternoon was our car and those of hundreds of other travelers like us, similarly numbed by speed and road miles. It seemed for a while that much of what is said about Kansas was true: endless flatlands, grasslands, cattle, and hay bales. As a midwesterner whose wife grew up on a farm, I saw bounty instead of boredom, but even so, after a while I wanted hills and a winding road. My midwestern niche, after all, is the Driftless Area, a rugged and bluff-filled region of the Upper Mississippi River valley.

We pulled off the road to spend the night in Emporia, a town of twenty-four thousand that is half agricultural hub and half university town, home to Emporia State University. We ate dinner (supper for true midwesterners) at a funky, youth-filled brew pub called Radius Brewing Company with colorful tap-brew names like Halfway to EveryBeer, but the downtown street names suggested a no-nonsense, practical alter ego: Merchant, Commercial, Mechanic, Market, and Exchange Streets, all laid out in a well-planned grid.

In the morning we left the interstate for Highway 50, a two-lane road that brought the landscape a little closer to our car windows. The Plains continued for another half hour, but then then the landscape started to dimple, lift, and fall away into valleys, subtly at first and then more pronounced as we turned onto State Highway 177, headed north, and spied the visitors' center to the Tallgrass Prairie National Preserve.

Like the parallel street grid of downtown Emporia, the state of Kansas is nearly rectangular, with a rumple on the northeast corner where the Missouri River dents the state boundary. Its four-hundred-mile east-west and two-hundred-mile north-south dimensions make it the fifteenth-largest state in the US.

When we left Emporia and began encountering the bubbling hills, we were entering the Flint Hills. The region at its widest stretches for ninety miles in east-central Kansas and extends 160 miles from near the northern border on south into Oklahoma, where it is typically called the Osage Hills.

The rolling hills are covered by 80 percent of the nation's remaining native tallgrass prairie, overlying a thin soil that in turn sits atop alternating layers of flint-pocked limestone and shale. But deeper beneath these are the remains of the mostly worn-away ancient Nemaha Mountain range that forms the backbone of the hills. The old mountain range is "like a hard pillow under a blanket," said Eric Patterson, lead park ranger at the Tallgrass Prairie National Preserve. The prairie and limestone/shale surface layers are like blankets swelled on top of this hard, ancient range, he added.[3]

One thing was for certain: Kansas certainly wasn't all flat. It wasn't all wheat and corn. And it wasn't all plain and practical.

Dianne and I pulled into the parking lot of the Tallgrass Prairie National Preserve around midmorning on an already blistering hot July day. The

prairie had already been racing across the landscape alongside our car windows for miles since we'd left Emporia. Grasslands stretched along the stream bottom across the road and on into the distant hills. Behind us, beyond the visitors' center, beyond the historical barn and outbuildings, prairie rolled up the hill through the cattle grazing areas and on into the Windmill Pasture where bison graze.

Prairie once covered four hundred thousand square miles of North America from the Rocky Mountains to the Mississippi River and beyond. Prior to the uplift of the Rocky Mountains, Eric explained, North America had a more homogenous landscape, "but the Rocky Mountains created a vast arid region on the lee side of the mountains." When the North American climate warmed and grew more even more arid 8,000–10,000 years ago, "that dryness created the environment where a new kind of plant could grow," namely prairie grasses and forbs. Before Euro-American settlement, 30 percent of the North American landscape was grass-covered.[4]

The Tallgrass Prairie National Preserve is home to bison and grazing cattle in eastern Kansas's Flint Hills. Prairie flowers and grasses flourish at the protected site.

North American prairie divides into three types along precipitation levels. Shortgrass prairie thrives in the driest climates (10–20 inches per year), particularly in the Rocky Mountain "shadow" stretching from southern Alberta and Saskatchewan to eastern New Mexico and northern Texas. Shortgrass prairie is typically ankle high and dominated by buffalo grass, blue gramma, and little bluestem.[5]

Tallgrass prairie, reaching from Manitoba to southeast Texas, requires 40–60 inches of rain per year. The tallgrass band typically reaches its eastern terminus at the Mississippi River, but it bulges eastward across the river through Illinois and northern Indiana, reaching as far eastward as Lake Michigan. Also known as the Eastern Prairie, it sports Indian grass, switch grass, and big bluestem, the lattermost of which can grow to be eight feet tall.[6]

In between lies a comparatively narrow band called the mixed-grass prairie, with features and plants of both the short and tallgrass varieties.

Kansas contains within it all three bands, with tallgrass beginning just west of the Flint Hills.

I didn't always love a prairie. Those maps of historical prairie ranges stopped just short of my home in Dubuque, Iowa, along the bluffs of the Mississippi River. My native landscape was the region of oak savanna, where prairie grasses spread out lawn-like beneath the hardwoods. In the modern absence of fire, those oak savannas turned to thick woods. I definitely had an eye for the woods.

Minnesota author Bill Holm differentiates the prairie eye and the woods eye in his essay, "Horizontal Grandeur": "A woods man looks at twenty miles of prairie and sees nothing but grass, but a prairie man looks at a square foot and sees a universe; ten or twenty flowers and grasses, heights, heads, colors, shades, configurations, bearded, rough, smooth, simple, elegant. When a cloud passes over the sun, colors shift, like a child's kaleidoscope."[7]

I had to learn to love the prairie.

Back at the preserve, Eric offered this advice to anyone who lacks the prairie eye: "Don't expect things to leap out at you. The prairie isn't one of those environments that will knock you over. It's more subtle than that. On the other hand, it often offers a vast scenery where you can see for ten or twenty miles."

But once you're done looking at the vista, look at what's in front of you. The Flint Hills sports seventy varieties of grasses and four hundred wildflowers and other forbs. Wildflowers take turns blooming from late spring through fall, each attracting their own set of pollinators.

For those of us whose idea of grass is the lawn we mow every Saturday, it can be a steep learning curve. "Your front lawn at home is the definition of a disturbed landscape," Eric lamented, "a monoculture of plants that have just one skill."

In fact, much of the skill of the prairie is unseen, underground. The root systems of prairie plants are vast, sinking ten to fifteen feet deep, enabling them to weather droughts and long, cold winters. Their thick, intertwining roots also hold the soil in place. "Three-fourths of the plants' biomass is underground," Eric explained. Above ground, the annual cycle of decay and regeneration of plant matter built up the midwestern topsoil, in some places up to eight feet deep.

Eric is definitely in love with the prairie. "Grasses are pretty resilient," he continued. "If it's dry, they simply slow their growth and store their energy in their roots. In wet periods, the grasses grow more full and tall." Prairie grasses are "the meek that have inherited the earth," he said. "They don't necessarily blow their own horn."

Indigenous peoples had long understood the prairies, hunting bison, elk, and other game and eventually cultivating corn, squash, and beans, and building cities in the midst of the grasslands that rivaled European cities in their time.

But European immigrants and their immediate descendants didn't initially grasp the significance of the prairie. "Many Europeans had grown up among trees, and thought trees were indicators of a healthy soil," Eric pointed out. An area that could grow trees could grow food, they thought, and could provide building materials as well.

Least Heat-Moon wrote that homesteaders brought with them the notion that "this hugely open spread was a kind of failed forest that needed only the hand of civilized man to redeem it from its appalling waste, and they reversed here their usual practice of axing wilderness: they planted trees" to overcome the open spaces.[8]

Eventually Euro-Americans discovered the fertility of prairie soils, though, and within a few decades uprooted and converted most of it to cropland. Today only 4 percent of the original prairie remains, and 80 percent of that total lies here in the Flint Hills. The Tallgrass Prairie National Preserve—coadministered by the US Park Service and the Nature Conservancy—protects and manages eleven thousand acres of the Flint Hills' 4 million prairie acres. But most of the Flint Hills' natural prairie lies in private ownership and is maintained by cattle doing the former work of bison, eating and trimming the prairie grasses and setting the stage for regrowth.

The proximity of limestone bedrock to the prairie grass surface differentiates the Flint Hills from much of Kansas's former prairie regions and is the main reason that so much of the prairie remains intact here. In the Flint Hills, limestone lies just inches below the soil and frequently protrudes, ensuring that much of the Flint Hills was never plowed, except in the stream bottoms where the soil covering was thicker.

Instead, the Flint Hills prairie has been cattle-grazed since the arrival of Euro-American ranchers, which, in the absence of bison, protects the prairie to a reasonable degree. "Ranchers recognize that they are dependent on the long-term health of the grass," Eric maintained. "If the grasses fail, so do they. And grasses do need to be grazed. Bison were the natural grazers, but now that task falls to cattle." Granted, there are differences between the grazing habits and behaviors of bison and cattle—with cattle

Bedrock close to the land surface protected eastern Kansas's prairie soils from the plow. As a result, 80 percent of the nation's remaining tallgrass prairie is found in this narrow band called the Flint Hills.

being rougher on the landscape. Bison, for example, might graze a section bare and then not return to it for several years, while ranchers typically confine cattle to limited, albeit large, acreages year after year. Ranchers simply have to "find the delicate balance between economic and ecological needs," Eric reasoned.

Dianne and I stopped briefly in the visitors' center to begin our annual reeducation regarding prairie wildflower and grass species' names, biology lessons that might last through the summer, but would definitely fade again

with the snuffing of the last flowers in autumn. At various times throughout the year, the prairie would host flowering splurges, showy evening primrose, coneflowers, wild bergamot, and more, each taking its turn across the seasons, racing across the prairie as red, orange, yellow, and purple buffaloes. We took photos of the sign boards, hoping to later identify what was currently in season.

Among the grasses, we could always identify big bluestem, as we were pretty much birthed among that six-foot-tall grass with its signature turkey-foot seed head in northeast Iowa. But for a while we could also pick out sideoats grama, Indian grass, hairy grama, and switchgrass, until those, too, faded from quick recall. Our short memories, though, have made each new season a fresh wonder of (re)learning.

Armed with ranger recommendations and a fresh supply of trail maps—Dianne collects these as if they were themselves prairie flowers—we headed out onto the preserve, following a gravel trail that, in non-COVID times, allowed buses to take visitors out to see the bison. The trail made no pretense of separating the hiker from the bison by fence or vehicle. "Keep your distance"—about one hundred yards—was all the warning given, with no particular suggestions about how to do that should a bull decide to charge. But the bison had apparently retreated to some tucked-away glen in the heat of this particular day. One bull was visible in the distance, lounging in the sun and then rising to eat, turn around, dust him himself off, and lounge again. We kept our eyes glued on him, out of wonder and respect, and to make sure that he stayed in place. What we would've done if he'd bolted, I have no idea. Dianne could probably outrun me.

But the trail did offer evidence of bison wandering. Here and there was a "buffalo wallow" formed by bison rolling in the grass to cool and dust themselves off. Their weight in the soft prairie soil creates a hollowed-out depression that in turn becomes a tiny mud pond in the next rainstorm, a haven to an array of smaller prairie creatures. In this totally secondhand manner, bison aided other life-forms on the prairie.

The herd is culled once a year, Heather explained, their numbers maintained in relation to the eleven hundred acres set aside for the bison. "Culling makes room for the babies," she said, adding that fourteen calves had been born in 2021. And it also makes room to bring in new stock, to reduce the risk of interbreeding. They send the culled bison to other refuges or to zoos, or sell them at market.

The grasses in July were thigh-high on the Tallgrass Prairie Preserve. This is the western, drier edge of the tallgrass prairie, and the grasses and forbs here wouldn't reach full height till late summer. Back home in the wetter climate of Iowa, they already towered over our heads. But by late fall, the Kansas grasses will reach four to five feet tall.

From the ridge top the shorter tallgrass, as it were, makes for impressive vistas. The land stretches out for miles, but only a scattering of barns and other outbuildings are visible in the vast horizon beyond the preserve. Nothing here but wind, the woods-eye person might say. But the wind was a Red Buffalo racing across the expanse, bending the grass tips in waves that rolled across the hills.

The stair-step ledges of the Flint Hills are quite apparent in the view from the ridge top. The ledges reflect alternating layers of limestone and shale formed in different sea-formation circumstances, with limestone deriving from shallow seas that teemed with shelled ocean creatures, and shale resulting when uplift put the sea bottom nearer a river delta where mud sediment compacted. Limestone is blocky, while shale is thinly layered, and where shale meets a downslope, it crumbles easily, undercutting the ledge of limestone above it. The pattern repeats, creating benches all the way down to where the hill meets the lowland.[9]

---

We left the Tallgrass Prairie National Preserve midafternoon to drive to nearby Council Grove, then on to our campsite at the Santa Fe Recreation Area on the Council Hill Grove Reservoir. The trees lining Fox Creek that paralleled the road for a few miles were the last we saw as Highway 177 steadily ascended into the treeless Flint Hills. We were still hot and dusty from our hikes; the hills looked even hotter and dustier from the comfort of the car.

Council Grove proved to be an immediate oasis. In this town of twenty-three hundred, we found Flint Hills Books, a recently opened bookstore in an elegantly renovated nineteenth-century storefront. We found the river walks hugging both sides of the Neosho River. We found our campsite. We found ice cream.

Council Grove has long been an oasis. The oaks and elms that grew along the Neosho River provided a natural haven for Native Americans amid the intense open prairie. Although the town itself was incorporated in 1858, its roots date back to the early 1820s when the newly minted Santa Fe Trail passed along today's Main Street.

An 1839 trail traveler described his natural refreshment at Council Grove before the town took root: "We stood at last beneath the sombre [*sic*] shadow of the old trees. We rode on through the thick wood, enjoying the grateful sensations occasioned by the transition from the burning heat of the prairie to the cooling shade of the grove."[10]

One of the oaks near the Neosho River gained fame as the Council Oak in 1825 when US government officials signed an agreement under its branches with the Osage tribe (and later the Kaw). The accord granted settlers safe passage along the 625-mile wagon freight trail that led from central Missouri to Santa Fe in what was then still part of Mexico. The signers named the agreement site Council Grove. The tree trunk is still preserved today on Main Street.

Preserved tree trunks are common in Council Grove. The trunk of the Post Office Oak still stands, protected from weather, along Main Street. The Post Office Oak contained a cache near the base of its trunk where trail travelers left messages for the next round of passersby. In addition, the preserved Custer Elm tree-trunk marks where General George Custer and his troops camped while patrolling the trail in 1867. Another burr oak stump stands at a business entrance with a plaque indicating its 1694 sprout date.

By the late 1840s, a small number of inns and supply stores had arisen amid the shade and waters of Council Grove as a "Last Chance" (the name of the earliest such store) to make purchases before continuing across the treeless prairie and into the southwest desert. In 1863 alone, over three thousand wagons and $40 million worth of goods passed through Council Grove.[11] This was another Red Buffalo, a slow one perhaps, but an ever-increasing swell of wagons rolling across the horizon.

Hezekiah Brake, traveling in 1858, worried about the upcoming weeks and months on the Santa Fe Trail during his stay at Seth Hayes's Last Chance Store and Inn: "On the last night before we started, the prospect seemed especially gloomy to me. Far away from my wife and child, and six hundred miles of constant danger in an uninhabited region was not a pleasant prospect for contemplation. But I laughed with the rest, joked about roasting our bacon with buffalo chips, and the enjoyment we would derive from the company of skeletons that would strew our pathway."[12]

The following morning, Brake sounded more optimistic upon leaving Council Grove: "We went off in grand style the next morning. The huge prairie schooner was well filled. We took with us for planting and feeding half a ton of shelled corn. Besides this, we had Hungarian-grass seed, rifles, boxes of crackers, bacon and sugar, robes, blankets, and many other articles—about two tons in all."[13]

Dianne and I wanted to know—in modern convenience, of course—what it must have felt and looked like to approach and leave the "last chance" of Council Grove on the Santa Fe Trail. First we drove eight miles east of Council Grove and, after a couple of misses, found the well-hidden but well-restored prairie at the Santa Fe Trail's Rock Creek crossing, where another supply store had set up shop in 1854. We hiked a quarter mile through the prairie to the creek. In 1865, Frank Stahl, hired by the US military to drive eleven hundred cattle across the prairie, lamented in his diary that one of his exhausted cows lay down at Rock Creek and he could not induce it to get back up. The next day, he and his men and the remaining cattle reached Council Grove.[14]

Returning to Council Grove ourselves, we stopped for ice cream (again) at the 1861-constructed Terwilliger Home Museum on the west side of town, the last home that travelers would have seen at that time before the final four hundred miles of their journey.

And then we drove west of town, exiting roughly along the same route as trail users would have used, leaving the last homes and businesses behind. A sign along a dusty road a few miles away pointed to wheel ruts and a twenty-foot swath made from the passing wagons, many of which carried two and a half tons of freight. But the heavily beaten path lay at a distance in a break in the trees on private property guarded by a barbed wire fence.

The fortunes of the Santa Fe Trail began to wane in the 1860s as the Kansas Pacific Railway became established. Junction City, KS, was a clear winner in the new developments, and an 1867 editorial in the *Junction City Union* proclaimed its victory over Council Grove: "A few years ago, the freighting wagons and oxen passing through Council Grove were counted by the thousands, the value of merchandise by millions. But the shriek of the iron horse has silenced the lowing of the panting ox, and the old trail looks desolate."[15]

The last gasp for the Santa Fe Trail came in 1880, when the Santa Fe Railway, a galloping, smoking, belching Red Buffalo, finally reached New Mexico, replacing the need for the wagon trail altogether.

But Council Grove shifted with the times. Rail service had come to Council Grove in 1869 with the Missouri, Kansas & Texas Railway (commonly known as the MKT, or Katy). Then came the Topeka, Salina & Western Railway in 1882. And finally there was the Council Grove, Osage City & Ottawa—later known as the Missouri Pacific—in 1886. Walking around town, we found Durland Park with its menagerie of depots original to the site or moved there, including one with a weathered façade announcing arrival at "M-K-T Council Grove, Kansas."

A century later, by 1994, the last of the railroads, too, had disappeared. But unlike many such rural midwestern towns, Council Grove didn't shrivel up with the disappearance of the railroad. Its population has remained remarkably stable since 1890, spurred on by restaurants, shops, and historical sites catering to visitors and campers from the nearby reservoir.

Back at the Santa Fe campground, we met our camping neighbor, who worked for the federal government helping local ranchers build ponds for cattle in the arid climate.

I thought of his ponds as Dianne and I pedaled out onto the Flint Hills Nature Trail on our bikes. The 117-mile crushed limestone trail, seventh-longest in the nation, stretches across east-central Kansas on the former railbed of the Missouri Pacific Railroad, running roughly parallel to the Santa Fe Trail. We picked up the trail in Council Grove and rode out into the dry, rocky countryside, rarely spotting house or human. Occasional ponds like those engineered by our camping neighbor offered refuge to small, scattered herds of cattle. The cattle seemed as surprised to see us as we them.

Patches of limestone poked through the prairie grass in surrounding fields. I thought of Ireland, of the limestone fields stretching across the Burren, but the shimmering heat on the Flint Hills Trail broke the illusion. Road cuts along the old railbed offered close-up observation of the geological story. Blocks of weathered limestone alternated with shards of shale and sharp-edged flint. One could make a tool from these, or pierce a bike tire so far from town.

We were a pair of Red Buffalo, riding hard into the wind for an hour or so.

Hot and dusty, we reluctantly set our limit, stopped, turned, and pedaled back to town.

Flint attracted the first humans to this region about twelve thousand years ago. A crystalline mineral of the quartz family that conglomerates in bands of limestone, flint can be chipped and flaked into sharp spear tips and other useful tools. Where limestone is exposed on a hillside, pods of flint may be found at or near the surface. The Flint Hills were markedly rich in such deposits.

Forty miles northeast of the Tallgrass Preserve lies the Claussen archaeological site, where habitations dating back 10,500 years and 1,200 years have been unearthed in close proximity. Among the commonalities found between the two sites separated by nearly ten thousand years were flint chards left over from toolmaking—flint that indigenous peoples had gathered from the local hills.[16]

By the 1800s, four main tribes lived in the Flint Hills: the Wichita, Pawnee, Osage, and the Kansa, now more commonly known as the Kaw. The Kaw, or the People of the South Wind, are most directly associated with Council Grove. The Kaw had been considerably weakened by the time of first contact with Euro-Americans, their population already reduced by 50 percent due to smallpox and cholera spreading from the east even before the Whites' arrival in Kansas. With pressures mounting in the advent of the Santa Fe Trail, the US government coerced the Kaw into the 1825 treaty exchanging land for promised annuities, and by 1846 they were further restricted to a two-hundred-fifty-thousand-acre reservation near Council Grove. In 1873, the Kaw were forced entirely out of Kansas to live on an even smaller reservation in Oklahoma—while the state had ironically adopted the Kansa name as its own.

The Kaw Chief Allegawaho complained bitterly, "Great Father, you treat my people like a flock of turkeys. You come into our dwelling places and scare us out. We fly over and alight on another stream, but no sooner do we get settled than again you come along and drive us farther and farther."[17]

Dianne and I stood near the banks of the Neosho and regarded the Kaw Mission State Historic Site, former home of the Methodist Episcopal Church's Kaw Mission School that operated from 1851 to 1854 with the goal of forced assimilation.

The tribe continued to be decimated by smallpox and other diseases, and by 1902, the federal government removed the Kaw from its list of federally recognized tribes, depriving them of what little assistance the government was willing to give.[18] However, the tribe was recognized again in 1959.

In recent years the Kaw have reestablished a presence in the Council Grove area. In 2002, it dedicated 168 acres of the Allegawaho Memorial Heritage Park it had purchased a few miles outside town. At this site in 2015, for the first time since 1873, the Kaw performed ceremonial dances in Kansas once again.[19]

Later I realized that Dianne and I had ridden our bikes past the ruins of the old Indian Agency building, just off the Flint Nature Trail as we returned to Council Grove. We took a short detour on the gravel road to learn about its restoration-in-process. We noted that the two-mile-long Kansa Heritage hiking trail lay just up the hill from the bike path. But we were hot, thirsty, and exhausted, and didn't go there. We had much yet to learn.

The Underground Railroad—another Red Buffalo, although this one running quietly, secretly, usually at night—passed through the northern Flint Hills in the mid-1800s. The hills begin to peter out by Wabaunsee County so that one encounters not a range of rolling hills, but occasional natural mounds, or monadnocks, popping out from an increasingly flat plains landscape.

It took some sleuthing, but Dianne and I found the Mount Mitchell Heritage Park on the northern edge of the Flint Hills down a dusty gravel road. Under a crisp, blue sky we climbed and circled the mound. We climbed amid stalks of purple hoary vervain and yellow compass plants, past a picnic spot halfway up the hill and past purplish-red boulders carried and dropped by ancient glaciers. At the top lies a Native American burial mound 1,000–2,000 years old.

The view from the top expands in all directions, although row crops and farmhouses are more frequent here than deeper in the Flint Hills. It would be a good promontory from which to watch for danger and to time one's passage.

The mound overlooks an old pathway of the Underground Railroad.

Kansas was birthed amid the hotbed of the mid-nineteenth-century slavery debate. The Kansas-Nebraska Act of 1854 established the Kansas and Nebraska Territories and directed them to decide for themselves whether to enter the Union as Free or Slave States. The next few years saw the onset of Bleeding Kansas, with armed proslavery groups from Missouri and armed antislavery groups from New England converging on the Kansas Territory to join in the general melee.

William Mitchell, from whom Mount Mitchell takes its name, was a Scottish-born immigrant who grew up in Connecticut and migrated to Kansas with the Beecher Bible and Rifle Colony, drawn by the opportunity to join the antislavery movement. He soon became leader of the Free State militia, the Wabaunsee Prairie Guards, that fought alongside abolitionist John Brown. Mitchell purchased a farm at the base of today's Mount Mitchell. A trail at the base of the mound served as a frequent route for escaped slaves, and Mitchell's log cabin served as a safe haven.

Over fifty people died during Bleeding Kansas's trial run of the American Civil War. After several attempts, an antislavery constitution was adopted by a two-to-one margin in 1859, and Kansas was admitted to the Union as a Free State in January 1861, less than three months before the Civil War began.

We circled the rim of Mount Mitchell, subdued in equal measures by the awe of the hillside prairie, by the plains landscape stretching in all directions, and by the lives and stories of those who had passed along its base.

The Red Buffalo had brought fire, but in its wake came new life.

We returned to the Tallgrass Prairie National Preserve a second day, this time taking the historical tour through the Spring Hill Ranch. In 1878, Stephen F. Jones purchased his first 160 acres along Fox Creek, arriving from Colorado with thirty-two train carloads of cattle.[20] Over time he acquired seven thousand acres at the Spring Hill Ranch and more cattle, and built an eleven-room, three-story Second Empire mansion with mansard roof and dormers that still stands on the preserve grounds, open to the public. Beyond the mansion lay several outbuildings, including a 110-foot-long, two-story barn built into the hillside to provide wagon access to both levels, a chicken coop with a vaulted ceiling, a curing house with portholes and a cupola for air circulation, and a three-seater outhouse (one seat lower for the children). All were constructed from limestone quarried on the property.[21]

These were good times for Chase County, where Jones's ranch was located. Between 1870 and 1880 the population had soared from 1,975 people to 6,081.[22] Jones donated land for a one-room community schoolhouse that opened in 1884 (and closed in 1930).[23] Today the schoolhouse sits alone amid prairie flowers, barely visible on the horizon from the Jones mansion.

Jones planted a large grove of trees on the swell west and north of the house, no doubt as a winter wind break and to guard against the blazing Kansas summer sun. Dianne and I hiked through the woods and down into the valley amid the punk-rock-purple hairdos of wild bergamot, brilliant orange butterfly milkweed, the five-lobed, pastel violet wood sorrel, and yellow cup plants and compass plants with descriptions we'd still remembered from the visitors' center. The proximity of limestone bedrock to the surface soils was apparent. On the hillside protruded Funston Limestone slabs that turn upright as they slowly slide downhill, earning the nickname of Tombstone Limestone. In the lowlands, white Eiss Limestone poked through the soil, round-edged and pocked with cavities due to erosion and dissolution. This was clearly untillable land.

Most puzzling were the thirty-plus miles of dry-stone fences capped with slanted, vertical coping stones that wound through the property. Had we been transported to western Ireland where stone fences like this still define the landscape? The treeless horizon might have said yes, but the dry prairie and the scorching heat said no.

Stone fences, we soon learned, are common throughout the Flint Hills. Early Kansas ranchers grazed their cattle in open ranges, herding them across the landscape as needed. As more settlers began arriving with the passage of the 1862 Homestead Act, pressures grew to end the open range. In 1870, the Kansas government began paying farmers to fence in their fields and pastures, and in the Flint Hills the most available fencing material was the limestone that protruded from the landscape. The farmers often employed Scottish masons who capped the walls with vertical coping stones, giving them a Celtic touch.

The population boom soon turned to bust. The Chase County population crested at 8,246 in 1900 and today stands at 2,744 (although it has seen almost 2 percent annual growth each year since 2018). Its population density today sits at 3.3 persons per square mile.[24] Ranching required ever more land for economy of scale. The average Chase County "farm" today is fifteen hundred acres, with ranches tipping the upper end of the scale.[25]

Establishing the Tallgrass Prairie National Preserve didn't go smoothly at first. Simultaneous with the downward economic and population trends came a growing interest among conservationists to preserve what little natural prairie remained in the American Midwest. To that end, in 1961, the National Park Service proposed a thirty-four-thousand-acre national park in the northern Flint Hills that soon became a fifty-seven-thousand-acre proposal with the option to purchase up to sixty thousand acres.[26] By 1979, the recommendation had grown to one hundred thousand acres, all of which could be acquired by eminent domain if not from willing sellers.[27]

For a rural Kansas culture already distrustful of the federal government, the proposal was outrageous. Here was another runaway Red Buffalo. Ranchers felt that their way of life would be threatened by so much parkland and that too much property would be removed from the tax rolls, most likely increasing property taxes for the remaining owners. In 1961, when Secretary of the Interior Stewart Udall dropped in by helicopter to inspect a potential property in the Flint Hills, he was greeted by the gun barrel of the rancher who was then leasing the land. The rancher became a "local legend for taking on the federal government."[28]

The federal government then changed its proposal from a national park to a series of prairie reserves, with properties obtainable only from willing sellers, but this mitigation was not much better received.[29] Della Wrae Blythe of the Kansas Grassroots Association didn't buy the compromise: "Preserve, reserve or whatever it's called, it's a park," adding that the Flint Hills would likely become either an "uninhabited no-man's land or a tourist trap complete with curio shops and hot dog stands."[30]

A solution began to take shape when the Spring Hill / Z Bar Ranch became available for purchase near Strong City, KS, with potential for a public-private collaborative ownership on an acreage much smaller than the early Park Service proposals. But local support still lagged at the outset. Ranchers held that if the federal government got even a foothold on the property, they would have the means to increase their presence.

After a number of iterations, a compromise—brokered in particular by Kansas senator Nancy Kassenbaum—resulted in the Tallgrass Prairie National Preserve. The preserve would be a public/private collaboration, with a nongovernmental entity owning 99 percent of the property and the National Park Service owning only the 180 acres on which sat the historic Jones family buildings, with prohibitions against purchasing additional property.[31] The Park Service, however, would staff the preserve and manage both the land and the herds of cattle and bison. This, finally, was a proposal that all could live with, and Congress established the National Preserve in 1996. In 2005, the Nature Conservancy stepped in as the private owner.

Back at the preserve's visitors' center, Heather Brown told us the short version of these events. "We're in a much better position with the community now that we've worked together," she said. Heather added that the Nature Conservancy continues to be a contributing member of the community by paying property taxes on the preserve, even though as a nonprofit organization they would not be required to do so.

Raising cattle keeps the preserve in touch with the community as well. Heather reiterated that cattle mimic the prairie maintenance of bison. While the surface bedrock made most of the Flint Hills unsuitable for tilling, cattle owners learned early on that the region was ideal for grazing cattle. In early times, as cowboys herded cattle across the open range toward markets in Kansas City, they discovered that cows increased in weight as they grazed across the Flint Hills prairie.

Today, the Flint Hills are an ideal location for finishing cattle by owners all across Kansas and beyond. Half a million or so yearlings are brought to the Flint Hills from April to September, before being sent on to market. Adding the cattle herds owned year-round by local ranchers, the Flint Hills are home to almost a million beef cows per year. Cattle that graze in the Flint Hills can gain one to two pounds per day throughout the summer before the rich nutritional value of the prairie grasses drops off by early fall.[32]

The Red Buffalo keeps the prairie alive, even for the cattle who have inherited the grasslands from the bison. Prescriptive burning of the prairie helps keep the grasses at their nutritional best and isn't just the purview of preserves and refuges. Private ranchers long ago learned the wisdom of Native Americans in burning the prairies to bring forth rich, spring grasses. "Prairie thatch blocks the sun from reaching the soil," Heather explained, "squelching off new growth." Fire puts the sun in contact with the land.

Eric Patterson added that grazing fields are most productively burned on a rotational basis. At the preserve, he said, rotational burning "mimics the randomness and patchiness of natural and indigenous fires."

I thought about the times when I've helped out with prairie burns back in Iowa. On one occasion I watched as a fire specialist set a line of flame into the wind at the edge of an intended burn site. The fire struggled against the wind, barely rising ankle high before petering out after ten feet. The fire-herder moved then to the far side of the field with his back to the wind and set another line of fire with his drip torch. Now the fire leaped like wild horses, roaring and crackling and raging orange and black. When it reached the blackened edge of the first burn, it dropped instantly to its knees, a trained and docile beast.

I was typically a gofer, appropriately entrusted only to a rubber flapper to tamp out any wayward flames. Once, though, an official handed me the drip torch and instructed me to lay out a line of fire across a field. I felt like I held the sun in my hands.

We drove again to beat the heat on our last day in Kansas, this time on the Flint Hills Scenic Byway. We joked about the state's utter practicality after spotting business signs like "The Tire Store" and recalling the literally named "Council Grove City Water Supply Lake." (No wonder there are so many Fundamentalists here, I said.) But we also found colorful, humorous, and artistic touches. There was the Screw My Nuts automotive repair shop in Bushong. And along the Flint Hills Byway there was the quixotic outdoor arts station near Matfield Green where we found whimsical, metal-sculpted wild stallions and a wooden Stonehenge to which one could, I suppose, harness a giant, imaginative horse.

There was a tension here between the tamed and the wild, a sense that the prairie had to be conquered but an equal sense that wildness gives the region life.

And then we were back on the highway again. Heat shimmered off the treeless, rolling hills. Along a sweeping curve and after a period of silence, Dianne finally said, "I need some trees," invoking our woods-eye heritage along the Upper Mississippi valley. Yes, I agreed, but added, somehow the vast open spaces are haunting and hallowed.

And then we were silent again, racing along the prairie like the Red Buffalo.

# 6

# Nebraska

## *The Niobrara River and Northern Sandhills: Drifting, Resisting, Rebuilding*

The Niobrara River was eating its bed.

The day before, we had been kayaking the Niobrara in northern Nebraska, occasionally dodging submerged boulders in the swift-moving, clear waters. And now we were standing on the riverbank a few miles downstream at the Norden Chute, watching its waters plunge fifteen feet over a lip of bedrock that marked its progress in downcutting its streambed. The river appeared to be draining from the lip of an oversized pitcher.

The Niobrara flows over five hundred miles from eastern Wyoming through the northernmost tier of counties in Nebraska before linking up with the Missouri River. Seventy-six miles of its midsection are preserved and protected as a National Scenic River, a designation that once split the community. Unlike most rivers on the Plains that wander listlessly through broad valleys, the Niobrara is a youthful stream that careens between tall bluffs on either side. Boulders in the riverbed are prizes it has licked away from the cliffs. In places, it has downcut three hundred feet into the surrounding upland plains. Locals have watched the Norden Chute creep

The swift-moving Niobrara River is still downcutting its bed in northern Nebraska. Long-time residents have watched the Norden Chute eat its way up the valley.

upstream over the course of a few decades, feeding its appetite as it clawed from the east side of the Norden Bridge to the west.

My wife Dianne and friend Dana had set up camp at Dryland Aquatics, a kayaking, canoeing, and tubing outfitter service located in the unincorporated town of Sparks, a few miles north of the river. Owners Ed and Louise

Heinert had "bought the town" in 1998 when they began operating the outfitting service. They run their business from the General Store, where local ranchers drop in for morning coffee and conversation.

Sparks consists of the General Store, Dryland Aquatics, the owners' family and their houses, and whomever might be camping on the grounds or sleeping in the bunkhouse prior to their river excursions. At the far end of the campground, an 1888-constructed community church harkens back to when the region was more populated than today. Euro-American settlers learned the hard way that the surrounding sandhills were more suitable for spread-out prairie cattle ranches than for small tillable farms.

But the entire river valley might have disappeared in the 1970s when the US Bureau of Reclamation proposed a dam that would have inundated twenty miles of Niobrara River valley, including the Norden Chute, to feed irrigation canals that would lead to outlying areas. Local opinion was divided—and still is—as to whether the dam should have been built, but an unlikely coalition of conservationists, canoe/kayak outfitters, and ranch owners successfully stopped the proposal.

Ed and Louise were among those who opposed the dam. "People back then didn't see the river as part of an ecosystem," Ed told us. "A lot of people didn't agree with us."

"But their grandkids might," added Dana.

---

Dianne and I had entered the southeast quadrant of Nebraska a few days earlier, after a visit to eastern Kansas. Almost immediately across the state border the landscape looked more like our Iowa home with gently rolling hills and endless rows of corn. Our eyes had grown accustomed to the sweeping, treeless hills and plains of northeast Kansas.

We were on our way to Lincoln to spend the day and night with our friends Dana and Graciela. Graciela was now the provost of Nebraska Wesleyan University but had previously been a Spanish professor and then academic dean at Clarke University in Dubuque, Iowa, where we live. Dana had been a professor of Spanish at Loras College, where I teach.

We'd been close to Dana and Graciela before their move and missed them dearly. Dana, a native Nebraskan, had taught me how to love a prairie. I had grown up among the hills of the Upper Mississippi valley and Driftless Area and had an eye for woods and sheer river bluffs. Back home, Dana had introduced me to the wide array of grasses and flowers in our small, restored Iowa prairies, had included me on expeditions to gather prairie seed, and—most cool of all—had invited me to prairie burns when he worked alongside county and state conservation officials. Plus, I got to hang out with his tan Labrador, Lilah, when we hiked.

Dana had another passion as well. In Dubuque, he'd worked with inner-city kids in a summer literacy program he'd started, called Future Talk. But he'd linked this literacy work with environmental education, bringing these kids and teenagers in contact with nature. While reading Aldo Leopold, he and his students built and set up birdhouses and cleared brush from overgrown trails and woodlots at nearby county and state parks. He took them camping and canoeing and kayaking on Mississippi backwaters. He was convinced that the Earth had stories to tell and that literacy included the ability to read the landscape.

Now moved back home to Nebraska, he'd continued with these related quests. In addition to his literacy work with Nebraska teenagers and young adults—many of them immigrants—he'd taken several groups kayaking on the Niobrara River, camping in the Sparks bunkhouse and getting them onto the river with Dryland Aquatics.

Dana had grown up thirty miles from the Niobrara in the farming community of Atkinson, Nebraska. Familiar with the area, he would be accompanying us for several days. He even made all the arrangements. Graciela couldn't make the trip with us, due to her work. Neither could Lilah, who was beginning to show her age and, besides, didn't like boats. So during the trip we received occasional updates from Graciela, usually that Lilah was still lying by the front door, waiting for Dana's return.

Dana, Dianne, and I left Lincoln, heading north and northwest. We crossed the Platte River and the Elkhorn, angling through the towns of Wahoo and Norfolk and Plainfield to reach Highway 20, and then headed west through northern Nebraska.

Most travelers encounter Nebraska along Interstate 80 in the southern third of the state. While Omaha, Lincoln, Grand Island, Kearney, and North Platte offer interesting metropolitan and small-town respites from the long drive, Interstate 80 mostly offers drawn-out miles (420 at its widest) across the Plains that fuel travelers' sighs. Too much corn in the east, too much wheat in the west, and everywhere too much sky. The Platte River offers some visual relief with a living, snaking band of green visible at a distance from the interstate where every road and row of corn lies straight and angular. Most travelers are unaware that the languid Platte River hosts half a million sandhill cranes each year during spring migration.

Heading north to the Niobrara, we crossed over the Platte where it swings north of Lincoln. For a while, the landscape on either side of the green divide looked the same. But Google Maps enthusiasts and the flyover crowd looking down from above would soon spot innumerable circular cornfields bathed by center-pivot irrigation systems in this most-irrigated state of the Union.

Slowly but noticeably, though, cropland gave way to grassland. The Plains began to dimple and lose its irrigation circles near Johnston in the north-central region of the state, within twenty miles of the South Dakota border. We were entering the northern limits of sandhill country. At first mere swells on the landscape, the sandhills soon grew to two-hundred- and three-hundred-foot mounds with tufted crests. They were grassed-over swirls of whipped cream. They were ocean waves frozen in the hundred-degree heat.

The sandhills of Nebraska cover nearly 25 percent of the state, mostly north of the Interstate 80 traffic flow and the Platte River. For 250 miles east-west and 100 miles north-south, they cover nearly 20,000 square miles of central and northern Nebraska, the largest sand dune region in the western hemisphere.[1]

Nebraska has been collecting sand for eons. Almost as quickly as the Rocky Mountains began uplifting in Wyoming, erosional forces began wearing them down, with ancient streams and rivers carrying dislodged and weathered sand and gravel into Nebraska and spreading it out across the plains through seasonal floods. As the Rockies rose taller, the lands in their eastern shadow grew drier, and the Nebraska plains transformed from forest to grassland. Ash from volcanic activity in Idaho added more soft layers to the sand base twelve million years ago.[2]

But the preponderance of Nebraska sand washed in and drifted in from the edges of the glaciers to the west and northwest a mere twelve thousand years ago. Then a drying, warming, postglacial climate intermittently left central and northern Nebraska sandhills exposed and desert-like. Relentless winds blew across the plains, as they still do today, and whipped the exposed sand into waves of crested dunes.[3] When moderate rains returned (central Nebraska currently receives about twenty-eight inches of annual precipitation), the dunes grassed over again and stabilized.

The waves were frozen in motion. As we drove alongside the emerging sandhills, some were small and rounded, while some stretched the lengths of football fields or more and showed gentle erosional creases. Some had towering edges, as if their waves were about to topple at the moment they grassed over.

Early Euro-American explorers were not smitten with the landscape. Gouverneur Kemble Warren wrote in 1856, "The scenery is exceedingly solitary, silent, and desolate, and depressing to one's spirits." And in 1866, geologist Ferdinand Hayden described the vegetation as "a few plants clinging with a sort of hopeless tenacity to the sides of the hills."[4]

That part of me attached to my Upper Mississippi valley's wooded hills might have agreed, but Dana had taught me how to love the prairie. It was a stark landscape, no doubt, hot in the July sun and barren of trees but nevertheless hauntingly beautiful.

Dana, Dianne, and I pulled into Dryland Aquatics in two cars about 4 p.m. on a July Sunday. Louise met us outside the General Store. Dana knows Louise and Ed from his previous trips with students. (For that matter, Dana is on a friendly basis within five minutes of meeting anyone.) He introduced us to Louise. Then they caught up on things: the impact of COVID on the outfitting business; Dana's plans to bring students in the early fall; our plans for the next morning's kayak trip. Louise said to watch out for her grandkids' puppy, still small enough to slip through the fence and dart out into the driveway.

We returned to our cars to drive into the park-and-tent-where-you-wish campground, but before we even started the ignitions, the pup sprinted across the lot in a beeline to Dana's car. Dana opened his door to greet the dog and flashed us a grin. "He's a dog magnet," Dianne laughed.

Or a magnet for all four-legged creatures. We picked out campsites near the edge of the property—we were the only tent campers on this Sunday evening—and soon a burro and two aging ponies came hobbling alongside the barbed wire fence where Dana commenced petting them. Dana had told us about them in an email as we were planning the trip: "My favorite citizen of Sparks is a burro. He bites, fights off mountain lions, and has a lot to say. We're kind of pals," he wrote.

The wind was relentless. Sparks lies on a plain on the northern edge of the Niobrara valley and the sandhills, so the wind, with no obstacles to block it, was pushing along an intense warm front. Dianne and I wrestled with our tent and watched it billow once it was in place. Dana set his up at a slight remove.

Down the road half a mile lay the Sparks Community Cemetery that Dianne and I explored after a cookstove dinner. Timeworn headstones with birth dates of 1834, 1827, and 1812, and death dates reaching as far back as 1892, sat among more recent burials. Some would have been pioneers who had settled in the Nebraska sandhills. Not a few were infants who died in their first year of life and young women who died during childbirth, attesting to the hardships faced by early prairie settlers.

Monday morning came gently to Sparks. The wind had settled, and an orange sun rose lazily alongside the community church. We cooked breakfast, brewed coffee, and wandered over to the General Store, where the morning's consultation had already begun. Dana introduced Dianne and me to Ed and to R, a local rancher who'd stopped in for coffee and conversation at the gathering table while his dog sat waiting for him outside in the bed of his pickup.

Ed had grown up not far away in South Dakota, and had been a high school teacher in nearby Valentine, Nebraska, for twenty-eight years. He'd taught history, humanities, and art until his retirement, whereupon he and Louise had started the outfitter business. Later I learned that several of the prairie paintings on the wall behind us were Ed's.

R was holding court this morning, though. The conversation was broad, not easy to pigeon-hole. It began somewhat predictably: upon learning that Dianne, like Ed, was a high school teacher, R intoned about the days when teachers could "strike fear into the hearts of misbehaving boys," as his teachers had with him. And how you couldn't get machine parts because truck drivers would rather collect COVID unemployment. But then it pivoted into how and why solar panels were superior to old-fashioned windmills for delivering water to ranch cattle, and about a National Public Radio program featuring a West African island that had been pioneering wind turbines for the last thirty years. And then to the pros and cons of cedar trees on the landscape. While cedars are native to the area, R maintained they had grown out of control and were a major factor in the 2012 wildfire that ravaged the area.

R's conversation and opinions were about as wide-ranging as the prairie.

Soon it was time to load up and head to the Niobrara. T, an employee, shuttled us down to the river. We dropped Dana's car at the Brewer Bridge, the take-out spot. Then T drove us and our kayaks up to the Cornell Bridge on the Fort Niobrara National Wildlife Refuge to launch our fifteen-mile float.

On the drive, we learned T's story. Originally from South Dakota, he'd gone to college out east and had recently moved near the Niobrara River to help out on his father-in-law's ranch. In addition to driving shuttle for Ed and Louise's outfitter service, he manages his online sales company where he markets a variety of duck-call hunting devices he'd invented.

His father-in-law's ranch, among many, had suffered greatly in the 2012 wildfire, which we were beginning to learn was part of the local historical lore. T also talked about the impact of Ted Turner bison ranches on the local economy. With Turner having paid top dollar for his ranches (five in northwest Nebraska), property values and property taxes had spiked for local ranchers, T complained. This, too, we would find to be a topic of frequent local talk.

T knew we were experienced kayakers but gave us a few tips about these particular boats and the Niobrara River. Beforehand I had wondered how the fifteen-mile float could be "rated" for four hours—back home such a distance would take all day. But when he launched me first into the Niobrara, I quickly understood as the river whisked me downstream ahead of the others.

The morning after kayaking I'd made plans to talk with Amanda Hefner, conservation assistant at the Nature Conservancy's fifty-six thousand-acre Niobrara Valley Preserve. Back at the campground, we sat in on the morning coffee conference again, this time with another rancher and a whole new range of topics, and then packed into our Honda and headed south to the Conservancy.

The Nature Conservancy entered the fray over the proposed dam on the Niobrara by purchasing two ranches along twenty-five miles of the river in 1980.[5] Its property straddles alternating banks of the river and also stretches to the northern reach of the Nebraska sandhills.

The idea of damming the Niobrara for irrigation and recreation had been brewing within the federal Bureau of Reclamation since the 1940s. It would be one of several proposed regional dams. In 1953, Congress gave "conditional authority" for the Bureau to move ahead with the Niobrara project, referred to as the O'Neill Unit. A decade later, plans began to emerge. The dam would, by means of canals radiating out from the pool, irrigate seventy-seven thousand acres of northeast Nebraska. Thirty thousand acres of privately owned land would be condemned for the project.[6] The lake would have been sixty feet deep and covered nineteen miles of river.[7]

Many locals were in favor of the dam, both for its irrigation purposes and for the potential to attract recreation dollars. Others were opposed, citing environmental reasons, private property rights, and protection of recreation dollars of a different sort from the growing reputation of the Niobrara as a canoeist's haven. And beyond all that, opponents pointed to the utter beauty of the landscape that would be flooded, and the loss of miles of clear, fast-moving river coursing at the base of some four-hundred-foot cliffs.

The plans were still in their infancy, but opponents were organizing. After Congress gave the go-ahead for construction of the dam, the Save the Niobrara River Association (SNRA) filed suit in federal court in 1975, arguing that the project's Environmental Impact Study (EIS) had been inadequate. In 1977, the US District Court halted construction plans until the EIS could be revisited.[8]

Nebraska politicians initially backed the proposal, but cracks in their line of support began to appear. In 1981, Nebraska congressman Doug

Northern Nebraska's Niobrara River flows swiftly over a boulder-strewn bedrock riverbed. A designated National Scenic River running amid sheer bluffs and numerous waterfalls, it is a haven for kayakers and canoers.

Bereuter was the first to break ranks. After the SNRA had successfully lobbied for a safety redesign that would have significantly boosted the project's cost and necessitated user fees for irrigation, Congressman Bereuter proposed killing the project on an economic basis. J. James Exon, who as

Nebraska governor had supported the dam, now as US senator changed his thinking and likewise opposed it.

In 1985, the federal government abandoned the plan, acknowledging that it did not have "sufficient support to warrant continued investigations."[9]

Today only one of the conceived dams is in place, the Merritt Reservoir on the Snake River nearly twenty-five miles away. We drove past it on one of our days in the area and watched the big boats and jet skis.

We entered the dusty gravel roads of the Nature Conservancy. The day's heat was already intensifying by 9 a.m., with a predicted high of 104 degrees Fahrenheit ahead of us. After a number of twists on the road and some rounded hills, we descended into the Niobrara valley. There's no gradual descent into the river valley—the plains just disappear into the earth, at the bottom of which the Niobrara is still cutting away at the bedrock.

Along the way we passed large groves of the charred remains of tree trunks from the 2012 wildfire and noted the progress of regrowth. We stopped for a while at the shoreline below the Norden Chute, mesmerized by the water pounding over the falls. I imagined the fate of a kayaker going over the edge. I imagined the path of a droplet as it went into free fall, smashed into the riverbed, thrashed in the undertow, and resurfaced to be hurtled downstream. I thought about the river eating its way from one side of the bridge to the other within a human lifetime.

We arrived at the offices of the Conservancy's Niobrara Valley Preserve (NVP) at about 9:30 a.m., and Amanda pulled in a few minutes later. Her dog Willa—named after Nebraska author Willa Cather—bounded alongside her and then dashed straight to Dana.

Amanda brought us to the recently constructed visitors' center, where another employee's dog greeted us as well. She ushered us into the community room, where in non-COVID times NVP made plans to host community events and educational groups. Long, sweeping views from a row of wide windows look out onto the preserve and the river valley.

Dana took a seat on the floor with Willa while Amanda sat down with Dianne and me at one of the community room's long tables. With a sweep of the hand, she pointed in the direction of the river valley. The dam, she said, would have been located just over there, near the Norden Bridge we'd just crossed.

"We're so lucky to have this place, thanks to all the people who worked to preserve it," she said.[10] On learning that we were staying at Dryland Aquatics, she told us that Louisa's father, Frank Egelhoff, had been among the most active preservationists, something that Louisa had been too modest to tell us herself.

"Other Nebraska rivers have been heavily impacted by crop agriculture," Amanda said, referring to both agricultural runoff and the effect of heavy irrigation in drawing down the river levels. "Up here, this river feels like a river is supposed to feel."

The NVP protects both sandhill prairie and forestland, a unique biological crossroads where eastern species meet the west. For example, this is the "farthest east range for ponderosa pines and the western edge for burr oak," said Amanda. Here the northern boreal forest blends with both eastern deciduous and Rocky Mountain pine and tallgrass, mixed grass, and sandhills prairie intermingle. Its fifty-six thousand acres represent one of the largest Nature Conservancy holdings in the country.[11]

And sandhill prairies, Amanda pointed out, harbor some specialized species not found in the east or west, such as sand bluestem with a root system even deeper than the acclaimed big bluestem of the tallgrass prairies. Its deep roots allow it to burrow into the sandy soil. "Anybody can love the mountains, but it takes a soul to love the prairie," Amanda quoted from Willa Cather.

The NVP leases land to local ranchers for cattle grazing, and through the 1980s and '90s acquired two bison herds totaling about one thousand head. They are very hands-off with the bison. They conduct a yearly roundup and sell off the oversupply, which helps with NVP's finances as well. "We want to be economically and environmentally sustainable," she explained.

To help heal wounds left over from the dam controversy, the NVP strives to be a contributing member of the local community. They pay property taxes on their grazing grounds by choice, not by legal requirement, as they are a nonprofit organization. In non-COVID times they offered their buildings to local meeting groups for free. They help ranchers with prescribed prairie burns and offer training sessions. "We have a very midwestern mindset," said Amanda: "Be humble, pay your taxes, and do good for the community." Adding to the community connections, Amanda pointed out, "Some people who have worked here at the Conservancy grew up in this area."

The NVP serves a broader community through research and educational programs, too. The preserve hosts everything from Future Farmers of America (FFA) and scout groups to university research teams and interns. The NVP's facilities truly facilitate this, with the large conference room, classrooms, kitchens, and lodging cabins.

The NVP shared in the community disaster of the 2012 wildfire as well. The fire, Amanda pointed out, ignited from lightning strikes in late July, during weather over one hundred degrees Fahrenheit and a drought (much like when we were visiting). The fire burned seventy thousand acres, including thirty thousand on the preserve.

The disaster "opened people's eyes to how we need to manage fire," Amanda pointed out. Explaining the role of cedar in spreading the fire, she said that while the prickly, combustible evergreen is native to the region, recurring fire—both natural and set by Native Americans—had historically kept the cedars in check. But fire suppression in modern times allowed cedars to spread. While ponderosa pines drop their lower limbs as they grow tall—which helps them resist wildfires—cedars keep their lower branches. When the lower branches ignite, it produces a "ladder effect" which takes the fire not only upward into the rest of the cedar but up into the ponderosa pine branches and the rest of the canopy, spreading fire throughout the woods as well as across the prairie.

Mature oaks—known for their resistance to fire—also burned, but soon sent up new sprouts from their roots, starting the renewal phase. At first the NVP waited to see if the pines would reestablish themselves, but eventually planted over two thousand new ones. "Mostly, though," Amanda said, "we are letting nature take its course. It gave us an opportunity to study the impact of climate change and wildfires." The Nature Conservancy obtained a grant to study the post-fire regeneration, collaborating with undergraduate interns. "Research showed that the rangeland benefited from wildfire," and the woodland cedars were reduced. "The oak savannas now look like oak savannas," she mused.

Amanda's energy and enthusiasm for the preserve is itself a little wildfire. Originally from central Nebraska, Amanda first came to the NVP as a student intern and has been here for six years as an employee. "I think it's the best part of the state," she beamed. "Such is the bane of the flyover states: there's so much more when you get off the interstate."

As we wrapped up our conversation, she gave us hiking maps for the preserve. "You encounter a little bit of magic when you descend into the Niobrara valley. Is this Middle Earth or the Middle Niobrara?"

Before we left, Dana made plans with Amanda to bring students to the preserve. We then drove a short distance to one of the two hiking trails on the preserve. By late morning the temperature was already well into the nineties.

It didn't take long before the crossroads of plant species and landforms were apparent. In the sandy soil grew big and little bluestem, leadplant, and yucca. The woods were graced with pine, cedar, birch, and oak. Charred remains from the 2012 wildfire demonstrated the ladder effect, where fire had consumed the cedar and climbed high into the pine branches. Sometimes there were signs of fresh growth, with pines so green and lush that their young, thin trunks could barely stand erect. But in small patches, the ground was still scarred and lifeless. Fire, Dana explained, is regenerative to the prairie and can be to the woods, but it can also blaze so hot in the woods that it sterilizes the soil for years to come.

The valley had more of an Iowa feel, with an oak savanna. Here the burr oaks kept a distance from each other but cast a canopy thick enough to discourage scrub trees on the forest floor. Prairie grasses grew lush beneath the oak limbs.

When we reemerged onto the sandhill prairie and returned to the car, we were soaked with sweat. Changing into shorts we began the daylong aversion—and revulsion—of removing ticks. We'd been forewarned and had taken all the precautions, but still found ourselves harboring plenty of the unfriendly arachnids.

Soon enough, Dana and Dianne launched their kayaks onto the Niobrara and caught up with me. The river skims briskly over its bedrock, sweeping the river bottom clean of siltation and debris. It's frequently punctuated with small rapids where boulders lay in wait just under the surface. I could usually float over them, but to exercise caution (and play at being adept with a paddle), I tried to dodge them. Small rapids keep you on your toes, so to speak, or at least perched in your seat and alert.

Even so, with the Norden Chute off-limits to boaters, kayaking the Niobrara doesn't require expert skills. But the river still commands attention more than most Midwest streams with its swift current and submerged boulders. *Backpacker* magazine has designated the Niobrara as one of the ten best canoeing rivers in the United States. More than eighty thousand paddlers and tubists visit the Niobrara each year.

The deep V of the river valley doesn't flatten out much into a flood plain, except slightly on the far side of the alternating bluffs. This is another sign of a youthful, still-downcutting river.

Groundwater weeps into the Niobrara from its bluffs at the junctures of geologic layers that are visually distinct by coloration. Waterfalls pour down, and a constant flow of streams refreshes it. They could drain the river at one mile, and by the next, it would have replenished itself.

The National Wild and Scenic Rivers Act went into law in 1968, with the Niobrara considered for inclusion even as there were plans to dam it. As the dam controversy neared its end in the mid-1980s, numerous landowners along the river—some say a majority—lobbied Senator Exon to seek further protection for the Niobrara as a National Scenic River. With the number of paddlers and tubists ever-increasing, landowners worried that uncontrolled development in the river valley would spoil its attractiveness. Already they were seeing an influx of new mobile homes.[12]

But other landowners along the river objected to its potential designation as a National Scenic River. The ranks of dam supporters and opponents reshuffled on the matter. Some who had argued against the dam also opposed the Scenic River designation, as they felt it would give federal control over their ability to earn income from their private lands.[13]

In 1985, Senator Exon introduced legislation for seventy-six miles of the Niobrara to be named a National Scenic River. The proposal languished for another five years because of divided positions among the Nebraska congressional delegation, but was reintroduced in 1990 after the dam project had been fully rejected.[14] The legislation assured that privately owned lands along the river would remain so, with protection for most existing grazing, farming, and campground activities in the valley.[15] But the National Park Service would have authority to regulate a half-mile of private property on either side of the river to keep the river corridor as natural as possible.[16]

The Niobrara received National Scenic River designation in 1991. But controversy occasionally resurfaces. In 1999, the Park Service put a two-year moratorium on new outfitters operating on the river, pleasing some and infuriating others. Concerns over alcohol use led the Park Service to prohibit alcohol on one portion of the river, largely solving the problem there but pushing it downstream.[17] A recent court battle was decided in favor of the Park Service in a dispute over how much of a property the federal government could regulate.[18]

The arguments were background noise to Dana, Dianne, and me as we plied the river. But we'd been learning more about the issues the past few days, and every once in a while it entered our conversation as we floated past another waterfall or cliff-side seep. Sometimes we paddled together and talked about the river, the valley, and the sandhills, and sometimes we spread out, each of us in our own space and thoughts.

The Niobrara's geologic story was written into the first bend in the river and then revised and edited with every succeeding turn. The youngest (topmost) layer in the bluffs is the Ash Hollow formation at 6–11 million years old, a sandstone that includes volcanic ash. Then comes the Valentine formation, a loosely consolidated sandstone deposited in ancient seas 12–13 million years ago. Drifting past cliff walls where this layer was close enough to touch, I reached out to find that it crumbled between my fingers.

Below these layers, and not far above the riverbed, lies the Rosebud formation, a siltstone laid down 25 million years ago. The northern range of the Ogallala Aquifer that feeds the nation's midsection from South Dakota to Texas sits atop the Rosebud formation in the Niobrara valley, and where it opens onto the river it gushes forth in waterfalls, seeps, and spring-fed streams. About 230 waterfalls tumble into the western half of the river.[19]

In our fifteen-mile stretch, we encountered Berry Falls, where a curved lip of water leaps twenty feet off a hardened Rosebud ledge and into the Niobrara. At Smith Falls, we pulled off river to eat a quick lunch and then hiked a quarter mile back into the park to watch the Ogallala discharging into a sixty-three-foot, one-bounce plunge that then raced down to the Niobrara. In between were numerous more cliff-seeps, smaller waterfalls, and streams fed by the Ogallala.

The bluffs themselves are rich in fossils, sorting them out by age, with early edition continental mammals on top, tropical wetland mammals and reptiles below, and sea creatures at the bottom.[20] The Ash Hollow, Valentine, and Rosebud layers have offered up the fossil bones of hedgehogs, mastodons, horses, and rhinoceroses, to name a few.

Another layer, Pierre Shale, sits below the Rosebud, rarely visible in the upper river but more prominent downstream where the river has cut even deeper. Here the fossils date 75 million years back to ancient seas, with marine reptiles, sharks, and bivalves.

This proliferation of Nebraska fossils isn't unique to the Niobrara valley. About a hundred miles southeast of Sparks sits the Ashfall Fossil Beds State Historical Park, a 360-acre active paleontological excavation site where fossilized mammals, birds, and reptiles continue to be unearthed. Twelve million years ago, Ashfall was a large regional pond that became an animal cemetery amid dense ash clouds in the aftermath of heavy volcanic activity one thousand miles away in Idaho. The volcanic ash plume drifted across today's Great Plains, but fossilized remains are especially concentrated at Ashfall, where the pond attracted agonized creatures whose lungs had filled with the abrasive dust. Perishing there, they fell into the ashen muck where their bones fossilized.[21] Due to continental drift, the geologic hot spot that created the Idaho volcanic activity now underlies Yellowstone National Park's renowned geysers in Wyoming.

The fossils show a snapshot of mammalian evolution. Over two hundred unearthed specimens include two- and three-toed horses that were precursors to the modern species. The menagerie also includes three types of camels, an early rhinoceros, and saber-toothed deer. Non-mammal fossils include early-version cranes, vultures, and turtles.[22] The site is sometimes referred to as a "prehistoric Pompeii."[23]

First discovered in 1971, today an eighteen thousand-square-foot enclosure called the Rhino Barn protects the site and offers shelter to the scientists and interns who continue unearthing fossils each summer.[24]

A drive through northern Nebraska is, I have insisted, more visually interesting than the well-traveled trek across the lower third on Interstate 80. Up north, you can lose any sight of human habitation for miles on end, even where the treeless landscape affords an exhaustive horizon.

With such a low population density, it's easy to forget that people have lived here at least nine thousand years, with the oldest identified site situated in the northwest corner of the state. In the Niobrara valley and northern sandhills, tangible evidence doesn't arise until the Plains Woodland period (AD 1–1000).[25]

The Ponca nation is most frequently associated with the Niobrara in historical times, although the Sioux were present in the western region of the river as well. The river's name is derived from the Ponca language, *ni obhantha ke*, or "spreading water river." A series of treaties, culminating in the 1868 Fort Laramie Treaty, led to the forced relocation of the Ponca to Oklahoma and the Sioux to what is now the Santee Sioux Nation reservation along the Missouri River.[26]

The US government established a series of forts at a distance from the reservation. One of these was Fort Niobrara, established in 1879, along the Niobrara River. The fort was relatively quiet. One soldier wrote home, "The most troublesome things here are the knats [*sic*] which walk all over a man and defy you to capture them. You can kill a mosquito, which is a consolation in itself, but a knat [*sic*] seems to have come into existence to keep the art of swearing alive." On the tragic side, Fort Niobrara was the launch site for troops engaged in the Pine Ridge massacre of 1890, just across the border in South Dakota.[27]

Euro-American settlement in the Niobrara valley accelerated in the late 1800s and early decades of the twentieth century, especially after passage of the 1904 Kincaid Act. The Kincaid Act expanded the Homestead Act's 160-acre allotment to settlers to 640 acres, acknowledging the impracticality of surviving on limited acres in the semiarid, sandy regions of western Nebraska. The expanded homesteading opportunity set off a new wave of settlement.[28]

Mari Sandoz describes the arriving settlers in her 1935 book, *Old Jules*. Covered wagons amassed at the local land office and then "vanished eastward over the level prairie. Many turned back at the first soft yellow chophills, pockmarked by blowouts and thwarted with soapweeds. Others kept on, through this protective border, into the broad valley region, with high hills reaching towards the whitish sky."[29]

The rush was short-lived, though, as homesteaders learned that the sandhills were not meant for row crops, and 640 acres was too small even for ranching in this climate. Over time, ranches consolidated into larger spreads, the average ranch today sprawling across 6,637 acres.[30] The population of Cherry County (home to Sparks, the Niobrara Valley Preserve, and the Fort Niobrara National Wildlife Refuge) peaked at 11,753 in 1920 and has steadily declined to 5,779 in 2019.[31]

But the economic impact of eighty thousand canoeists, kayakers, and tube-floaters is significant. After ranching/agriculture, health and educational services, and the retail industry, tourism ranks next in employment in Cherry County. The average floater spends fifty-five dollars a day on direct expenses and another forty-six dollars in restaurants and retails shops.[32] The results are evident in Valentine, the county seat, with its well-stocked bookstore, and western clothing and home décor shops. There are at least six hotels/lodges, many of them quite new, impressive in this town of twenty-eight hundred.

But back on the road, in the countryside, the reality hits again in a sweeping glance across the horizon: the population density of Cherry County is approximately one person per square mile.

The unpeopled landscape could be encountered practically anywhere off the main highways and the river valley, but we encountered it profoundly in the Valentine National Wildlife Refuge thirty miles south of Valentine. Scattered windmills providing water for cattle as they are herded from field to field among the sandhills to prevent overgrazing are the only indication of human presence in the Valentine Refuge.

The more surprising feature of the refuge is the preponderance of lakes and wetlands in this dry climate. Unlike the lakes and wetlands of the northern Midwest that are largely fed from snowmelt and rain, the sandhill lakes are mostly recharged from beneath, as the Ogallala Aquifer tops out barely beneath ground level, and the lowlands dip into the water table.[33]

The seventy-two-thousand-acre refuge was established in 1935 for migratory birds and other wildlife. Wildlife inventories have indicated 270 species of birds, 59 species of mammals, and 22 of amphibians and reptiles. About thirteen thousand acres of the refuge are wetlands.[34]

In the over one-hundred-degree-Fahrenheit heat, we decided on a tour by car. We drove the ten-mile Wildlife Tour's narrow gravel road, stopping at each interpretive site and anywhere else Dana could read us a story in the landscape. Tallgrass prairie grew where the water table is nearer the surface, and mixed-grass prairie and prickly pear cactus in the higher, drier elevations. Sand blowouts marked where high winds have ripped away most of the cover vegetation. The grasslands housed prairie chickens, grouse, sandpipers, hawks, and golden eagles. The occasional woodlands hosted bluebirds, buntings, and warblers.

We stopped to watch the wind dimpling across Dewey Lake and wished we had brought our kayaks. I imagined migratory birds descending on the lake in fall and spring: trumpeter swans, canvasbacks, ruddy ducks, mallards, herons, pelicans, and sandhill cranes. The refuge was designated as a Globally Important Bird Area in 2001.

Dana found an intact deer skeleton next to the lake and held it up for photo-ops, no doubt to be sent home to Lilah.

We drove down another unmarked, one-lane, gravel- and rock-strewn road as well, testing the limits of our Honda CR-V as we bounced along. The route wandered through a treeless bottomland prairie. Occasional cattle grids suggested an abandoned ranch. We turned around in a barely wide enough spot in the road when the threats to our car's suspension, plus the brooding weight of our isolation, overtook our curiosity.

From there, we drove forty-five miles back to camp, watching the sun set on a herd of bison on the upland plain.

Fort Niobrara was closed as a military post in 1906 but reopened in 1912 as the Niobrara Big Game Reservation, and was later renamed the Fort Niobrara National Wildlife Refuge. Its founding mission was to be a "breeding ground for native birds." Today the refuge comprises nineteen thousand acres.[35]

We pulled into the Fort Niobrara Visitors' Center and offices on our last full day in Nebraska to talk with Matt Sprenger, the wildlife refuge manager. Matt greeted us at the door and brought us into his office. Fort Niobrara, he explained, was one of the earliest national wildlife refuges. The conservation movement was still relatively new to the country and had captured the public's imagination. The National Audubon Society was still freshly established (1905), and influential figures like President Teddy

Roosevelt wanted to preserve game for hunting. "These were people who had seen the wilderness, and then had seen it disappear," said Matt.[36] The former fort's prairies thus became a hunter's paradise for wild birds and small mammals. Hunting is still allowed in specific areas of the refuge, observing Nebraska state regulations.

Bison and elk were reintroduced to the refuge soon after its inception. But in keeping with the refuge's founding mission, the big game served to keep the prairie healthy for bird habitat. That, of course, doesn't stop big game from stealing the show.

The 350 head of bison live generally free from human interaction but are occasionally culled to prevent overgrazing or traded with other refuges to increase genetic diversity.

One of the main interactions with the bison is moving them from their winter range in the wilderness area north of the Niobrara River to their summer range to the south, herding them across the river twice a year on the Buffalo Bridge. "The older animals seem to know that in September or October we'll move them into the wilderness area," Matt mused, adding that the "females usually lead the way."

Back on the refuge's roads, we drove past the corrals where bison and the refuge's Texas longhorn cattle are occasionally rounded up. In November 2011, the longhorn cattle herd was culled and volunteer cowboys undertook a 180-mile, old-fashioned cattle drive—including through the streets of Valentine—to deliver them to a new home at Fort Robinson State Park in northwest Nebraska.[37]

We drove to an overlook of the Niobrara valley and then hiked a short loop past another waterfall and down to the river. A small group of kayakers silently paddled past us as we watched from the woods.

Back in the car again, we approached the bison herd loitering beside and on the refuge road. We crept in their direction for a closer look and lingered at a short distance for a while. But not wanting to part the sea of bison with our car—for our sake or theirs—we turned around.

After launching our kayaks, we counted off the bridges crossing the Niobrara so as to not overshoot our take-out point, beginning with our put-in at the Cornell Bridge at the western edge of the Fort Niobrara National Wildlife Refuge. Soon we slid under the Buffalo Bridge. The 1921-constructed, steel-truss Berry Bridge was the first of three bridges on the National Register of Historic Places that we'd encounter on the float. Next up was the gleaming, silver 1903 Allen Bridge. The Verdigre Bridge, built in

1910, was moved 155 miles to Smith Falls in 1996, where it is now semiretired as a footbridge crossing the river. Last up would be the Brewer Bridge, built in 1899 and moved to its current location in 1921.[38]

We'd visited the Brewer Bridge on our first night camping at Dryland Aquatics, walked halfway across it, and contemplated the swift-moving Niobrara. Sighting it on the river from our kayaks after a ninety-degree bend, we made a beeline, one by one, to the north shore and pulled our boats ashore.

Then we drove out of the valley and back up into the Nebraska plains and left the Niobrara to its hidden work of eating through the sandhills.

# 7

# Missouri

## *The Katy Trail: Down Through the Layers*

*Sunday 21st Septr. 1806*

*At 4 P M we arived in Sight of St. Charles, the party rejoiced at the Sight of this [hospitable] village . . . we observed a number of Gentlemen and ladies walking on the bank, we Saluted the Village by three rounds from our blunderbuts and the Small arms of the party, and landed near the lower part of the town.*

—CAPTAIN WILLIAM CLARK,
UPON THE RETURN OF CORPS
OF DISCOVERY TO ST. LOUIS.[1]

Dianne and I crossed north over the Missouri River, riding the bike lane that hangs as a ledge from the side of the steel-girded Highway 54/63 bridge at Jefferson City, Missouri. Love locks adorned the chain-link fence on the river side of the lane. Cars and semis rushed past on the main bridge. Below us, the river frothed with waves whipped up in the west breeze. By the time it reached Jefferson City, the Missouri had already completed its descent from Montana's mountains and the Dakotas, and was now halfway through its final, mostly eastward sweep through the state that carries its name.

A bicycle ramp known as the Corkscrew connects a spur of the Katy Trail to a bridge over the Missouri River leading to Missouri's capital, Jefferson City.

We were headed from Jefferson City to the two-mile spur connecting to the Katy Trail, a 240-mile crushed limestone bikeway that rides on the bed of the defunct Missouri-Kansas-Texas (MKT, or Katy) Railroad and bisects much of the state of Missouri.

This was my second venture to the Katy Trail, Dianne's first. I'd been there a few years earlier with a group of guys I bicycle with. We'd rented a stately home in Jefferson City for a three-day stay. When we met the landlord at the property, she told us the residence had previously been a funeral home. That prompted a lot of juvenile humor among six men who had somehow passed through late middle age without having conquered adolescence. Any more embalming fluid in the cooler? Would you like some Cream-mate with your coffee?

I remembered the trip fondly. I recalled the two-hundred-foot, yellow-white, sheer limestone bluffs edging the trail near Rocheport, with the Missouri River lapping on the other side. And riding through the train tunnel at Rocheport, and then exploring its adjacent wildlife refuges. Having lunch and a beer and listening to live music along the river at Cooper's Landing.

And I'd remembered, too, the bike lane across the Missouri River at Jefferson City, especially at its end, the structure called the "corkscrew," a rectangular, three-level spiraling ramp where cyclists descend or ascend between the bridge and the trail spur. Riding the layers of the corkscrew was like passing over the same ground on the backs of successive stories.

I'd wanted to return to the Katy Trail for some time with Dianne. I'd wanted to ride down through the spiraled layers again and add some new ones as well.

## Missouri River

The story starts, of course, with the Missouri River, or at least this is where I pick it up. Missouri River enthusiasts and those of us who live instead along the Mississippi compete over some basic facts. Most sources say the Missouri is the longest river in the United States. Thus states the National Park Service and the US Geological Survey.[2] Calvin Fremling, in his time-honored, Mississippi-focused *Immortal River* claims the Missouri River is 2,340 miles long and the Mississippi River 2,301 miles.[3] But *Britannica* and the online digital mapping system ArcGis give a slight edge to the Mississippi.[4] Some put it more diplomatically: the Missouri-Mississippi river system, together, constitutes the fourth longest in the world and has the fourth-largest watershed, proving there is power in collaboration.[5]

Truth is, it's hard to measure the length of a river.

Truth is, they are both mighty long rivers.

Truth is, the Missouri River as it runs alongside the Katy Trail looks a lot like the Upper Mississippi as it passes by my home in Dubuque, Iowa.

As with the Mississippi, the Missouri River story starts with glaciation. Pre-Illinoian glaciers—the oldest—ordained the river's east-west passage across today's state of Missouri. The river set its course at the base of the glacier and carried away its meltwater. Running deep and strong through successive glacial period melt backs, the Missouri downcut through the limestone bluffs that parallel the river along the Katy Trail.

But that older river did not drain down from the north. Montana and Dakota streams once coursed northeastward to the Hudson Bay, not southward to the Missouri. The more recent glaciers of the Wisconsinan Age detoured these northern waters down through the central Dakotas and then along today's Nebraska-Iowa border and the upper third of Missouri's western border until they joined up with the old, original Missouri River channel. This southerly path is more recently carved and steeper than the older riverbed that winds more leisurely through Missouri.[6]

*Monday May 14th 1804*

*Rained the forepart of the day . . . I Set out at 4 oClock P.M. in the presence of many of the Neighbouring inhabitents, and proceeded on under a jentle brease up the Missourie.*

*—Captain William Clark*

## Machens

During the winter of 1803–1804, Captains Meriwether Lewis and William Clark were making final preparations for the Corps of Discovery voyage that President Thomas Jefferson had commissioned to explore the newly acquired lands of the Louisiana Purchase.[7] Clark and the crew of over forty men were hunkered down at Camp Dubois at the confluence of the Wood River and the Mississippi a few miles north of St. Louis, training for their various roles and readying the keelboat and two pirogues for the journey ahead.[8] Lewis split his time between camp and the home of the prominent St. Louis businessman, Pierre Chouteau, gathering the last of needed supplies.

With Lewis still in the city, Clark and the crew set off from Camp Dubois late in the day on May 14 after a rainy morning, coasted a few miles down the Mississippi, and entered the Missouri River. The rain that had delayed their start returned in the afternoon.

Clark and the crew camped their first night on a bend in the river only a few miles upstream on the Missouri. It would take another day and night's encampment before they reached St. Charles to rendezvous with Lewis. Moving upriver was arduous. The crew was aided by sails when the winds cooperated but generally propelled the boats upstream against the current by pivoting into the riverbed with long poles.

The Katy Trail's eastern terminus at Machens lies less than a mile inland from the Corps' first night's camp. The trailhead isn't particularly noteworthy. The Katy simply starts (or ends) there, largely unannounced, at its junction with a backroad turnoff from State Highway 94. Off-road parking accommodates only a few cars. That's OK. The unofficial, more popular terminus is twelve miles southwest at St. Charles.

Machens marked the western terminus of the MKT. From there, it joined forces with the Chicago, Burlington & Quincy (CBQ) Railroad, entering St. Louis on the narrow isthmus between the Missouri and Mississippi Rivers.

Midway between Machens and St. Charles once lay the town of Black Walnut, settled in 1810 by German and French-Creole immigrants. William Schnare operated a steamboat landing there in the 1850s. Schnare's yellow Lab would howl to alert his owner to the arrival of steamboats, earning the port its nickname of Yellow Dog Landing.

When we arrived at the Black Walnut trailside station on our bikes and took a few minutes' rest, another yellow dog greeted us, circled a few times, lapped from a water dish left for him along the path, and then disappeared into the cornfields in the direction of the river.

*May 16th Wednesday 1804*

*We arrived at St. Charles at 12 oClock a number Spectators french & Indians flocked to the bank to See the party. This Village is about one mile in length, Situated on the North Side of the Missourie at the foot of a hill from which it takes its name Petiete Coete or the Little hill This village Contns. About 100 houses, the most of them Small and indefferent and about 450 inhabitents Chiefly French, those people appear pore, polite & harmonious.*

*—William Clark*

## St. Charles

Clark and the crew spent five days in St. Charles awaiting Lewis's arrival from St. Louis.

The wait must have seemed long to a crew ready to embark, and the temptations irresistible at the last town the Corps would see until their return. During the stay, three men were charged as Absent Without Leave (AWOL) for tarrying too long at local establishments. One of them was additionally charged for behaving in an "unbecoming manner" at a dance. The men were court-martialed and the lattermost whipped.

The crew attended Mass at St. Charles Borromeo Church on Sunday morning, and Lewis arrived by nightfall with an entourage of St. Louis's finest, including Pierre Chouteau's half-brother, Auguste Chouteau.[9]

The Corps set off up the Missouri River the very next day.

Dianne and I stood on the bank of the Missouri at St. Charles straddling our bikes. It was a clear, blue day. The river was neither racing nor sluggishly passing along the grassy shore. "Deliberate" might put it best, due to the river's size. Clark, ever the cartographer, had measured the river's width during his stay at St. Charles and found it to be 720 yards.[10] But the river ran wider and shallower back then. Today the lower Missouri is channeled into a narrower, deeper, and faster flow to stabilize its path and accommodate barge traffic.

St. Charles is lovely, on and off the bikeway. Coming in from Machens, the trail passes some industrial sections on the north before arriving at Frontier Park with its larger-than-life statue of Lewis, Clark, and Lewis's dog Seaman, who became a valuable and loved member of the Corps of Discovery. The town of fifty-six thousand today is a northwest outer suburb of St. Louis. Its origins lie with the French trapper Louis Blanchette, who located his operations twenty-seven river miles west of the confluence in 1769. When others began settling nearby, he called the aspiring village Les Petites Cotes, or the Little Hills.[11] Reflecting the complicated bureaucracy of the Louisiana Territory—still culturally French but officially ceded to Spain—the town briefly adopted the Spanish name of its first Catholic church in the 1790s, San Carlos, which was soon anglicized to St. Charles. The town took on new prominence as the first state capital in 1821 when Missouri officially became a state. The designation was intentionally temporary until the newly plotted, and more centrally located, Jefferson City was well enough established to take on the governmental role farther west along the Missouri River.

Back on the bikes, we explored St. Charles for a while. Carefully navigating the historic brick streets on skinny road bike tires, we found the first capitol. Nineteenth-century brick storefronts with scrolling facades lined Main Street. We stopped at the reconstructed 1893 Victorian-style Katy Railroad depot. We somehow missed the modern recreation of the St. Charles Borromeo log church. We ate at the Bike Stop Outpost and Café, debating between the Lewis and Clark Wrap and the Katy Sandwich among other menu items, and checking out the new bikes, rental bikes, and cycling equipment. It's a bit of a disease.

Then we headed southwest out of St. Charles on the Katy Trail. For a while the route was industrial and urban, winding past construction sites, a cement factory, a casino, and a multipurpose event center with a sprawling parking lot. Then the city gave way to country at Greens Bottom, a marshy lobe where the river looped southward. The trail hugged the hills at the far north of the marsh. The Missouri has likely shifted course through the marsh many times. Greens Chute, a thin, natural canal, probably marks an earlier river path. The MKT Railroad chose a bluff-side route at the edge of the river's wanderings.

Still, the MKT flooded, and often. The last time the tracks flooded was in 1986. They were never used again.

*Monday 21st May 1804*

*Set out [from St. Charles] at half passed three oClock under three Cheers from the gentlemen on the bank. . . . Soon after we Set ut to day a hard Wind from the W.SW accompanied with a hard rain, which lasted with Short intervals all night*

*—Captain William Clark*

Less than four years after the much-celebrated Transcontinental Railroad laid its final spike connecting east and west, the MKT Railroad pushed south from Junction City, Kansas, through "Indian Territory" (known as Oklahoma after 1907), and reached the Texas border on Christmas Day 1872. It reached Dallas in 1886, Houston in 1893, and San Antonio in 1901.[12] By way of new and existing tracks, the MKT simultaneously expanded eastward through Missouri, from Kansas City on the west and through Clinton and Sedalia to the south. It completed its Missouri sweep by linking to St. Louis's other rail lines at Machens in 1894.

The company's fortunes peaked and plummeted several times, and included various mergers and alternative rail routes. In the 1890s, the MKT was awash in cash, but by 1915 its fortunes crashed, and it was placed in receivership. It turned around again, reaching new heights of prosperity during World War II. But by the 1960s it was losing money. It ceased operations through most of Missouri in 1986, and in 1989, the MKT was legally dissolved.[13]

Almost immediately, Missourians, led by financier Edward Jones, began planning to convert the railbed into a bicycle trail. Construction began in 1987, and the first section of trail opened at Rocheport in 1990. In that same year, the Katy Trail became a Missouri State Park.

In 1991, the Union Pacific Railroad donated the remaining thirty-three miles of right-of-way from Sedalia to Clinton. The final segment of the Katy Trail, from St. Charles to Machens, was opened in 2011.

But the conversion to the Katy Trail was not without controversy. At its inception, the MKT had obtained easements rather than purchases for the land its rails would run on. Since the easement agreements had stipulated "for a railroad, and for no other purposes," adjacent farmers through the decades had expected that if the railroad ever ceased operations, the easements would end, and the land rights would revert to themselves or to their successors.[14]

Landowners were surprised to learn that a bike path was being planned for the Katy route. They felt blindsided upon learning that a 1983 amendment to the National Trails System Act of 1968 had created a work-around, establishing "rail banks" that allowed trails to be built on railbeds until such time—if ever—railroads would return to the route.[15] A decade-long lawsuit was resolved in 2002 with a federal court deciding that the government was within its rights to build a trail, but not without compensating the landowners. Compensation of $410,000 was awarded to thirteen landowners, with lawsuits pending from nearly three hundred other landowners.[16]

Meanwhile, the Missouri Department of Natural Resources estimates that four hundred thousand annual Katy visitors bring $18 million in yearly revenue to shops, restaurants, and hoteliers along the trail.[17]

> *May 24th Thursday 1804*
>
> *Above this Isld is a Verry bad part of the river, . . . The Swiftness of the Current wheeled the boat, Broke our Toe rope, and was nearly over Setting the boat, all hand jumped out on the upper Side and bore on that Side untill the Sand washed from under the boat and wheeled on the next bank by the time She wheeled a 3rd Time got a rope fast to her Stern and by the means of Swimmers was Carred to Shore . . . This place I call the retragrade bend as we were obliged to fall back 2 miles*
>
> —*Captain William Clark.*

## Marthasville

Our bicycling friends Darwin Hill and Penny Splinter set off on a six-day, 454-mile ride on the Katy Trail in late April 2021.[18] They started from Marthasville, ten miles southwest of St. Charles. Darwin is a riding machine, logging at least ten thousand miles every year. Penny does about half this mileage, still surpassing Dianne and me. Their ride took them to the southwest terminus in Clinton and back, and included some side trails as well.

It's always fun listening to Darwin and Penny recall their trip, due to their distinctly different riding styles. Penny remembers the scenery with tall cliffs and the Missouri River, and the quaint cabins they stayed in at Windsor, near the end of the trail. Near Windsor, Penny recalls the clip-clop of Amish horses' hooves on the nearby road. Were she to ride the trail again, Penny would explore the towns more. She'd stay in the same place for two nights in a row, and ride east one day and west the next, covering only as many miles each day as she'd like. Then she'd move on for another two-day stay farther down the trail.

Darwin's trip is about speed and miles. He tells us about the two 100-mile and the two 70-mile days at a 15–16 miles per hour pace on the crushed limestone surface. He rather laments the two days of only covering fifty miles. His vision for a next visit would be to skip the accommodations entirely and ride completely self-contained, as he has done many times on other trips back home in Wisconsin.

Even so, with clothing, equipment, and a few snacks, Darwin was carrying thirty-five pounds for himself and Penny on the Katy Trail. "Next time," he joked, "you can carry the bags."

Penny shot him a look. "Oh, so I can ride slower?"

We've ridden with Penny. She doesn't ride slowly.

*May 27th Sunday 1804*

*As we were pushing off this Morning two Canoos Loaded with fur &c. Came to from the Mahars nation, which place they had left two months, at about 10 oClock 4 Cajaux or rafts loaded with furs and peltres came too one from the Paunees, the other from Grand Osage, they informed nothing of Consequence, passed a Creek on the Lbd Side Called ash Creek 20 yds wide, passed the upper point of a large Island on the Stbd Side back of which Comes in three Creeks one Called Orter Creek, her the men we left hunting Came in we camped on a Willow Island in the mouth of Gasconnade River. George Shannon Killed a Deer this evening*

*—Captain William Clark*

## Hermann

Just downstream from the mouth of the Gasconade River lies Hermann, MO. We'd been fighting some serious headwinds on the trail before we decided to head into Hermann for lunch. The town lies two miles distant and across the river from the Katy Trail. The bridge across the river is handsome with black guard rails and a protected bike lane.

Some towns along the Katy have withered over the last several decades, a fate not uncommon in rural America. We passed towns with gray, boarded-up mills, abandoned houses, and houses in disrepair and in need of paint or siding. But as soon as we crossed the bridge and entered Market Street, we knew that this town of twenty-four hundred people was alive. There was the Hermann Wine Trail tasting house, representing seven local wineries. The Deutschheim State Historic Site preserves the home of one of the town's German-immigrant founders. Tourist shops and restaurants lined Market Street and the downtown: Liberty Glass Works stained glass studio; Ricky's Chocolate Box; Grape Expectations Guest Haus; the Hermann Wurst Haus; and the Doxie Slush, a slushy bar named after the owners' dachshunds. We ate at the Downtown Deli & Custard Shoppe, chasing our sandwiches with generous helpings of custard ice cream.

The nation's longest bicycle trail, the Katy Trail, runs through central Missouri along the Missouri River before dipping to the southwest. It follows the path of the defunct Missouri-Kansas-Texas Railroad, colloquially known as the M-K-T or Katy Railroad.

Hermann was founded by German immigrants who first settled in the eastern US but were looking for land in the rural Midwest. In 1836, the German Settlement Society of Philadelphia purchased eleven thousand acres along the Missouri River and sold shares to other eastern-settled German immigrants to establish the "heart of German-America where the customs, language, and traditions of the fatherland could be translated into a New World setting."[19] A prosperous wine country, the region came to be known as the Little Rhineland on the Missouri.

Melissa Lensing, director of the Hermann Area Chamber of Commerce, said that Hermann's modern success is due to entrepreneurs who "have taken out-of-the-box ideas and created things that people are drawn to."[20] Some of the business owners grew up in Hermann, but others were visitors who fell in love with the town and decided to bring their own creative business ideas. Melissa herself is one of those outsiders who fell in love, and not just with the town: "I married a Hermann German," she joked.

Special celebrations like Oktoberfest and Maifest bring particularly large gatherings to Hermann, but tourism is strong throughout much of the year. Wineries have been a longtime draw, along with a brewery and four distilleries. Visitors enjoy the authentic German architecture. It helps that Hermann is a stop on the Amtrak route that runs between St. Louis and Kansas City. But the Katy Trail brings visitors, too. "We have bicyclists here every day," Melissa told us.

Like other small towns, Hermann has had its ups and downs. "Before Prohibition, the wine industry was the focus of Hermann. Prohibition was devastating to the town," said Melissa. Federal officials "came in and burned down the vineyards and poured out the wine."

Only a few years ago, a lot of buildings were up for sale. But today over 140 guest houses, B&B rentals, and boutique hotels offer more than four hundred guest rooms right within Hermann. And with no major chain hotels, "it's not your cookie-cutter experience," Melissa added.

Dianne and I were drawn to the old world architecture and historical museums. My journalist roots attracted me to the Carl Strehly house, where Strehly and Eduard Muehl began publishing the *Hermanner Wochenblatt* (the *Hermann Weekly Letter*) in 1845, the first German and antislavery newspaper west of the Mississippi.

Abolitionist sentiments were common in Hermann from the start. Melissa pointed out that the immigrants had left Germany because of tyranny and weren't about to tolerate oppression in their new home. Hermann residents frequently took in escaped slaves and ushered them north to freedom.

This small town still actively supports human rights. Its 2021 Black History Month agenda included a symposium titled "The Shared History of Germans and African Americans in Missouri"; offered German music and jazz; and put on a production of the play, "German Abolitionists of Missouri."[21]

After lunch we pedaled back to the car parked along the Katy Trail. A rainstorm blew in from the west, so we explored the rest of Hermann and its hilly countryside by Honda. Our wipers could barely keep up with the downpour.

*May 30th Wednesday 1804*

*Rained all last night Set out at 6 oClock after a heavy Shower, and proceeded on, passed a large Island a Creek opposit on the St. Side just abov a Cave Called Monbrun Tavern & River, passed a Creek on the Lbd. Side Call Rush Creek at 4 Miles Several Showers of rain the Current Verry Swift river riseing fast*

*—Captain William Clark*

## Portland

Dianne and I made a brief car stop at Portland, MO, near the Tavern River, where the trailside parking lot opened to a Missouri River access. The unincorporated town began in 1831 as a riverside port for local agricultural products. With several bends on the river on either side of the town, its long, straight shoreline made it a good stopping point for steamboats to load and unload merchandise. Steamboats ran on the Missouri from 1819 until the early 1900s, until they were replaced by barges and the MKT Railroad, both of which made fewer stops, sealing the fate of small river towns like Portland.

We talked with a through-rider at Portland. He'd started in Kansas City, pedaled sixty miles on the Rock Island Trail to its junction with the Katy, and now was headed for St. Louis. He fished in his saddlebags for some trail mix as we talked. But his bags looked curiously light for such a trip: a few changes of clothes, light rain gear, some nonperishable foods, a bedroll. No tent, as he slept under the stars. He didn't talk long. He was only halfway across the state.

We got back in the car and drove to our next stop.

A Confederate flag flapped in a farmhouse yard just outside of town, a reminder of a more sinister past and present.

*June Sunday 3rd 1804*

*Capt Lewis & George Drewyer went out & Killed a Deer, We Set out at 5 oClock P M Cloudy & rain, West 5 Ms. to the mo. of Murrow Creek Lb Sd. a pt. St. Side Keeping along the Lbd Side 1 Ms., passed the mouth of a Creek on Lbd Side 3 ms., I call Cupboard, Creek, mouths behind a rock which projects into the river, Camped in the mouth of the Creek aforesaid, at the mouth of this Creek I saw much fresh Signs of Indians, haveing Crossed 2 Deer Killed to day. I have a verry Sore Throat, & am Tormented with Musquetors & Small ticks*

*—Captain William Clark*

## Jefferson City

Just past the mouth of the winding Moreau River on the opposite bank from the Katy Trail lies Jefferson City. Out on the trail we could see the Missouri State capitol gleaming from two and a half miles away, across marshes, sandy fields, and the river. We found the spur into town, rode the ascending layers of the corkscrew ramp, and biked across the Missouri River.

With a population of forty-three thousand, Jefferson City hasn't succumbed to the bureaucratic and financial sprawl endemic to many state capitals. Its downtown is smartly revitalized with coffee shops, small restaurants, and Irish pubs. Bike trails wind through the city's hills, and the streets are low volume and bicycle friendly.

The city streets part around the capitol like the river around an island. The sprawling white structure was constructed in 1917 after two previous capitols had burned.

On my bike in front of the capitol, I couldn't shake from my mind the fact that Missouri had been the Midwest's only slave state. When the Confederate states left the Union, though, Missouri found itself divided. Elected in a pro-Southern wave in 1860, Governor Clairborne Fox Jackson found himself at odds with a closely divided legislature that ultimately voted to remain in the Union. An irate Governor Jackson set up a shadow government in Boonville dedicated to the Confederate cause, while federal troops soon occupied Jefferson City for the duration of the war.

One of the city trails leads past Lincoln University, established at the end of the Civil War by members of the Sixty-Second and Sixty-Fifth US "Colored Infantry" for the benefit of freed African Americans. Known then as the Lincoln Institute, it became part of the Missouri state university system in 1879, and today enrolls about eighteen hundred students.[22]

The only member of the Corps of Discovery crew who hadn't freely chosen to join was York, a slave to Captain William Clark. York is mentioned over fifty times in their journals for his prowess in hunting (being enslaved, he would not have been allowed a gun in ordinary circumstances before the journey), his swimming abilities that frequently helped to rescue or stabilize the boats, and for other contributions. He and the Native American guide Sacajawea were permitted to "vote" on crew decisions. But out west, York was also paraded and put on display before the Nez Perce to both entice their curiosity and incite their fear, as they had never before seen a Black man.[23]

After the two-year odyssey ended, York appealed to Clark for his own freedom, or at least to be hired out in Louisville, where his wife was enslaved. Despite his exemplary work on the expedition, Clark kept him in bondage for several years thereafter. At one point Clark became so aggravated by York's requests that he had his former crew member whipped.

When Clark finally released York, he gave him a wagon and several horses, with which he established a business hauling goods.[24]

> *Wednesday the 6th of June 1804*
>
> *Mended our Mast this morning &, Set out at 7 oClock under a jentle breise from S. E. by S passed the large Island, and a Creek Called Split rock Creek at 5 ms. on the S. S. psd. a place to the rock from which 20 yds we. this Creek takes its name, a projecting rock with a hole thro a point of the rock, at 8 ms. passed the mouth of a Creek Called Saline . . . the water excessivly Strong, So much So that we Camped Sooner than the usial time to waite for the pirogue, The banks are falling in Verry much to day river rose last night a foot. . . . I am Still verry unwell with a Sore throat & head ake.*
>
> —*Captain William Clark*

## Cooper's Landing

We resumed our ride at Cooper's Landing, a boat dock, bar, and campground located near the site of the 1820s up-and-coming shipping town of Nashville that washed away in an 1844 flood. There was no live music on the September weekday we visited, but the Missouri Monsters were next in the lineup. Even so, it was a great place to grab a cold beverage and relax. We sat for a while on Adirondack chairs in front of the store and watched the river pass. The current is frisky here. Down at the river's edge, a signboard tells of the Missouri's swift flow around this corner where the inside curve holds eight feet of undisturbed sand while the outside bend has been scoured nearly to bedrock. A series of wing dams protect the shoreline on the outside curve.

From there, the trail passes by Boat-henge, a whimsical display of a half-dozen 1950s-era boats half sunken into the ground in a graceful arc, their alternating bows and sterns protruding, creating a bit of local kitsch in homage to the river.

## Providence and the Columbia Spur

Only a boat ramp access to Perche Creek remains of the town of Providence, a Missouri River port in the mid-1800s. The port mostly served the landlocked city of Columbia that was growing exponentially in the decades after the University of Missouri was established there in 1839 (the first public university west of the Mississippi). Enterprising merchants financed a ten-mile plank road connecting the inland city with Providence. A local lumber mill sawed enough lumber to build a mile of road per month until its completion in 1856. But the plank road was rickety (one traveler wrote that his wagon rose and fell on the tipsy boards "like a ship riding ocean waves"), and in the frequently wet climate the planks quickly deteriorated. Eventually Columbia was connected to railroads, including a spur to the MKT, and the need for a river port declined. What's more, the river's fickle channel frequently shifted after major floods, and the final injury to Providence came when the channel shifted westward, leaving the town with only a shallow creek connecting it to the river.[25]

At the town of McBaine (population seventeen in 2020), we detoured two miles west and north of the Katy to see a four-hundred-year-old burr oak. When Lewis and Clark's crew passed this way, in view of the tree, the oak would already have been two hundred years old. Damaged in a lightning strike, the tree is now girdled with lightning rods to protect it from future storms. After four hundred years, a little support is warranted.

Near McBaine, the rail spur that once connected Columbia with the MKT is now a bicycle trail linking the university town with the Katy. This is where our friends Eric Eller and Neely Farren-Eller began their five-day, 270-mile ride in July 2021, starting from Columbia and heading 120 miles east on the Katy Trail to Defiance, and 120 miles back west again to Rocheport before returning to McBaine and Columbia.[26]

The couple, who both teach with me at Loras College in Dubuque, were no strangers to the area. Eric grew up in Boonville, the next town west of Rocheport. As teenagers when the railroad was still operating, he and his friends would hike through the Rocheport Tunnel and explore the caves in the bluff that Lewis and Clark had noted in their journals.

Neely was the original cycling enthusiast of the two, however. Neely has been riding since she was twelve years old, exploring country roads near her family's Ohio farm. In winter, she set up an old bike trainer in a barn loft. Her workouts were so intense that the tires spinning against the trainer's resistance cylinders "sounded like a freight train," she said. She kept her cadence high: "Sometimes it was so cold in the barn I could see steam rising off me." Later, when she taught at Michigan Tech, she'd ride the twenty-two-mile roundtrip to school and home, even in the winter.

When Eric and Neely set out from Columbia, there'd been a foot of rain the week before, and it was still coming down. They carried saddlebags of 30–40 pounds and planned on tent-camping. Because of the weather, the trail was mushy, with wheel ruts and washouts, making the riding slow and unsteady. "I'd come to a near standstill in the ruts, and not be able to get out of my clips, and fall over," Neely admitted.

"But you were very good at falling," Eric laughed.

They had four flats and ruined three tires, mostly because of the trail conditions after the heavy spring rains. Neely remembered thinking at one point, "'I don't know if I can do one more second,' but you know you have to go on."

In the end, the weather didn't bring them down. They camped at the Bluffton Barn, about ten miles west of Hermann, owned by a "good ol' Missouri bachelor guy" who was one of the early proponents of the Katy Trail. Nearly exhausted, they called him from the trail to make arrangements to camp on the grounds. Doug, the owner, offered to pick them up, but Neely said no, they'd ride in. But they welcomed showers, dinner, and breakfast at the Barn.

They finished their overnights at a bed-and-breakfast in Rocheport. They enjoyed the quiet of the trail. "And we met so many interesting people," Neely said. They talked with a man cross-country racing from Portland, Oregon, to Washington, DC, against twelve other cyclists.

On this leg of the trip, Dianne and I met the most fellow travelers, too. At the Columbia spur we met three women who had ridden the trail west to east three times, and now were training for their fourth trip, this time east to west, against the wind. They'd stayed at such unusual places, they told us, including a converted silo, a converted bank, and—you guessed it—a converted funeral home in Jefferson City. We met a biologist, who shooed us around a copperhead snake sunning on the trail. And we met Brad Landolt, owner of Rocheport's Meriwether Café and Bike Shop.

*June 7th Thursday 1804*

*Set out early passed the head of the Island opposit which we Camped last night, and brackfast at the Mouth of a large Creek on the S. S. Of 30 yds wide Called big Monetou. . . . A Short distance above the mouth of this Creek, is Several Courious Paintings and Carveing in the projecting rock of Limestone inlade with white red & blue flint, of a verry good quallity, the Indians have taken of this flint great quantities*

*—Captain William Clark*

## Rocheport

Wildlife refuges line up back-to-back in the miles leading to Rocheport: the Overton Bottoms Conservation Area, the Big Muddy National Fish and

Wildlife Refuge, and, just beyond Rocheport, the Davisdale Conservation Area. Limestone bluffs close in on the river here, though somewhat obscured by tree foliage. Along the Eagle Bluffs Conservation Area, the river sweeps away to the west and south, the distance between the trail and the river now lined with tree-filled marshes. Soon enough, we'd passed Petite Saline Creek and The Hole in the Rock that Lewis and Clark mentioned in their journals.

In the seven trail miles leading to Rocheport, though, the Missouri River returned trailside, and the two-hundred-foot sheer bluffs rose directly above us, shorn of tree cover. Raptors hung in the looming, yellow-white rock walls. When they swooped out over the river, their shadows reeled across the trail and across our backs.

But they'd find no carrion at Rocheport, another small town to which the Katy Trail had brought new life. With a tiny population of 201, it seemed to boast as many restaurants, antique stores, wineries, and general businesses as people. We stopped for lunch at the Meriwether Café and Bike Shop, where we ate outside along with a dozen other riders and visitors. I ordered the Meriburger ("Add a Clark, make it a double"!).

I talked with owner Brad Landolt, who came to the nearby University of Missouri-Columbia twenty years ago as a student and stayed on.[27] I was again curious that some communities embraced the trail while others seemed at best to ignore it.

Landolt, who's owned the Meriwether Café and Bike Shop since 2017, said, "There's a collective will that exists in some of these small towns to really lean into the trail. It's been a great thing for the town. But it does take effort and planning. Even twenty years ago, the first time I came to Rocheport, it was distinctly different." The town's been cleaned up a lot, Brad explained, through such measures as an ordinance prohibiting trailers on residential lots.

"There had always been a trailside business at the Meriwether location since the trail opened in 1990," Brad said. At first it was just a trailer that sold hot dogs and brats. It closed briefly in 2015 after its owner retired but was reopened and upgraded by some Rocheport locals who "saw themselves highly invested in the success of the town," but were not restaurateurs. They hired Brad to run the café. Brad bought it shortly thereafter and reopened it as the Meriwether Café and Bike Shop.

"I wanted to create an atmosphere that feels like an extension of the parklike setting of the trail," he explained. The menu consists mostly of made-from-scratch dishes. "I'm proud of our relationship with a lot of local producers."

Brad was adamant that the Rocheport area offers the best of the Katy Trail. "There are so many really high-quality sights in this short section that people even just coming for a day ride can see much of the best of what the Katy Trail has to offer. We have the only railroad tunnel on the trail, some of the best views of the bluff and the river, Native American petroglyphs, and several Lewis and Clark historical sites."

From the Meriwether Café & Bike Shop, Dianne and I got back on the trail and rode toward the Rocheport Tunnel. Before entering the tunnel, though, we walked our bikes down a side path to the Dana Bend Conservation Area. The Missouri twists away to the west again here, while the trail continues in a northwest slant. Although Dana Bend was largely dried up after the summer drought that had followed the spring rains, it serves as an important wetland recharge station for migrating waterfowl when seasonal rains return.

The 243-foot train tunnel was blasted through the bedrock in 1893, destroying the petroglyphs that Clark had described. The northwest entrance leading to Rocheport sports a handsomely arched limestone face, while the entrance leading away from the town opens into the gaping natural rock. Inside, the tunnel is perfectly arched in brick, cut stone, and natural rock. The manicured northwest opening may have heralded the arrival at Rocheport, which, at the time of construction, was still near the peak population of 823 it had achieved by 1870.

Beyond the tunnel, the trail was broken up with ruts, potholes, and large chunks of rock that had worked their way up from the softened trail bed. The heavy spring rains had damaged the trail, and this section had not yet been restored. A trail bridge was out of commission not far ahead, detouring riders onto nearby roads.

But our time was up, so we reversed course and headed back again. We'd miss the southward turn of the Trail as it crosses the Missouri and leaves the river behind. We'd save Boonville for another trip, another town that has "leaned into" the Katy Trail. We'd miss the sweep through Sedalia and Windsor and on through to the southern terminus at Clinton, MO. But we'd still managed to ride 120 miles over a few short days on the Katy Trail.

Heading east and southeast again on the Trail, we came back up through the layers: the Missouri River, Native Americans, Lewis & Clark, Euro-American settlement, slavery, abolition, the Civil War, the MKT Railroad, the Katy Trail, and ourselves on our bikes on a glorious September afternoon.

The Corps of Discovery would continue up the Missouri River along today's Nebraska-Iowa border, swing through South and North Dakota and on past the river's headwaters in Montana, cross the Rocky Mountains through Idaho, and descend the Columbia River between Oregon and Washington. They'd arrive at the Pacific Ocean on November 15, 1805, and begin their return journey. All told, they'd log eight thousand miles in twenty-eight months.

Many in St. Louis had never expected to see them return.

> *Saturday 20th Septr. 1806*
>
> *[On the return trip, nearing a village about seventy-two river miles west of St. Charles.]*
>
> *We passed the enterance of the Gasconnade river below which we met a perogue with 5 french men bound to the Osarge Gd. village. the party being extreemly anxious to get down ply their ores very well, we Saw Some cows on the bank which was a joyfull Sight to the party and Caused a Shout to be raised for joy at ____ P M we Came in Sight of the little french Village . . . the men raised a Shout and Sprung upon their ores and we soon landed opposit to the Village. our party requested to be permited to fire off their Guns which was alowed & they discharged 3 rounds with a harty Cheer*
>
> *—Captain William Clark*

## Thursday, 16th September 2021

After loading up the bikes, Dianne and I drove to St. Louis, crossed the Mississippi, and followed the river home to Dubuque.

# 8

# Illinois

## *Cahokia and the Upper Mississippi Shoreline: The View from the Center*

From the platform top of Monks Mound, I can see seven miles across the Mississippi River to the St. Louis Gateway Arch. You might say I'm seeing nine hundred years into the future.

Monks Mound is the largest of about seventy remaining Native American mounds at the Cahokia State Historic Site near Collinsville, IL, not far from the east bank of the Mississippi. A United Nations Educational, Scientific, and Cultural Organization (UNESCO) World Heritage Site, Cahokia was a Native American city with as many as twenty thousand residents by AD 1100, larger than London at that time. No American city surpassed its size until 1800.

The city once encompassed six square miles, including at least 120 mounds of varying types, and was the hub for numerous outlying villages as well.

By AD 1300, Cahokia had disappeared.

I live alongside the Mississippi River in Dubuque, Iowa, and see it almost every day. The view from my fifth-floor, east-facing office window at Loras College overlooks a mile and a half of downtown Dubuque, the Mississippi river, and the Wisconsin shoreline bluffs. Dubuque lies across the river from the Wisconsin-Illinois border, so when I walk south from my home almost every morning to the nearby bluffs, go bicycling through the Mines of Spain, or greet the sunrise at the Julien Dubuque monument, I am looking at the Mississippi River as it first passes along the Illinois shoreline. In summer when we kayak on the river, we skim the river's back. On a hot day I will dip a baseball cap into the flow and drape it across my head, letting the Mississippi drip slowly from my hair.

I have lived along the river for almost all of my sixty-plus years, except for graduate school and a couple of ventures in Ireland. The longer I live here, the more I fathom the pull of the river through time. Generations and civilizations have come and gone along its banks, and the river was here before that, too.

Dianne and I had been meaning to visit Cahokia for several years, this late, great pre-Columbian civilization near the Mississippi River in today's western Illinois. But one thing and then another had conspired against our going there until recently. In the meantime, we notched our way down along the Illinois shoreline, hiking, bicycling, and kayaking. Eventually we stitched together enough experience and memory to line the Mississippi shoreline from East Dubuque to Cahokia.

The Mississippi River forms four hundred miles of the Illinois western border from East Dubuque to East St. Louis, not far from Cahokia, and another 105 miles from East St. Louis to Cairo at Illinois's southern tip at the Ohio River. It is the four hundred northern miles that I have come to know best about Illinois.

## Casper Bluff

Dianne and I found Casper Bluff, an eighty-five-acre preserve five miles southwest of Galena along the bluffs of the Mississippi River, one long-ago spring. From our vantage two hundred feet above the river, we looked down at hundreds of white gulls skimming above the river in a slow, silent glide.[1] The evenness of their flight was interrupted by a rolling breeze on which the gulls alternately crested and then rode down. The flock looked like a shook white blanket flapping in slow time.

Up on the bluff, the oak savanna also swelled and undulated from ravine to hilltop. Some of the swells were human-made, though. They were Late Woodland Culture Native American burial mounds dating back to AD 600.

About twenty visible mounds, including Illinois's sole remaining intact thunderbird effigy, grace Casper Bluff. The restored oak prairie savanna looks much like it must have a thousand years ago.

The mounds at Casper Bluff have long drawn the attention of archaeologists. In the late 1800s, amateur archaeologist William Nickerson noted thirty-eight linear mounds, twelve conical mounds, and one thunderbird effigy with a wingspan of at least one hundred feet. In the intervening years, some mounds were lost to pioneer plowing, and some have simply slumped away with time.

The Casper Bluff mounds are a remnant of the six hundred mounds once documented in Jo Daviess County. Archaeologist Phil Millhouse pointed out that the area's Native American population—dating back nearly twelve thousand years—was growing in the early centuries of the first millennium AD. Settlements along rivers like the Mississippi were numerous.

Mound building was important to the Woodland culture communities. Mound artifacts, Millhouse noted, dispel any notion that the Woodland culture was primitive. "Woodland peoples had artifacts that spanned the continent," said Millhouse, including "Gulf Coast conch shells, sharks' teeth, Lake Superior copper, galena, obsidian from Yellowstone, and a host of other exotic cherts (flint) and other materials."

In sacred landscapes, nature is sacramental, full of signs. On the spring morning of our first encounter with Casper Bluff, there were no road or city sounds, no whining of boats on the river. Above the river, hundreds of gulls skimmed along on a crest of wind. And on the bluff, the thunderbird effigy rode a crest of oak savanna.

## Mississippi Palisades State Park

Our connections with Mississippi Palisades State Park near Savanna, IL, go back longer than I can remember. It's one of the first places where we cross-country-skied shortly after we married in the early 1980s. We took our kids camping there when they were young. I took students there. The rock structures keep us tethered to this place.

Sentinel Rock is a two-hundred-foot limestone tower cleft from the sheer-bluff river overlook at Mississippi Palisades.[2] The rock pillar is split as if by a lightning bolt halfway to its base, and a boulder half the size of my car seems perilously wedged in the crevice. It looks as if it could have tumbled and lodged there during a recent rainstorm.

But appearances are deceiving. The forces that shaped the park were slow and powerful.

Ancient seas laid down these thick limestone beds from the compressed shells of sea creatures 450 million years ago, interspersed with muddy shales and occasional seashore sandstones. When the land heaved up out of the sea, a great river bisected the new continent. The Mississippi ran brim to brim between its bounding bluffs when the final glaciers melted just twelve thousand years ago and carved steep, rock-walled borders to the swelling rush. When the meltwater subsided and the river dropped to its current bed, river bluffs such as the Mississippi Palisades testified to the great surge.

At Ozzie's Point, the northernmost overlook, we found a man with a tripod camera focusing on a distant freight train crawling northward at the base of the bluff in wide, sweeping S-curves. The train slowly achieved one hidden turn and then another, and soon enough pounded the tracks two hundred feet beneath the sheer limestone cliff face. In an instant, the train quickly chuffed beneath the rock tower, its engine shifting octaves in Doppler fashion.

At Louis Point, a couple sitting on a bench gave friendly nods as we approached and watched the haunting sight of a dozen turkey vultures swooping in graceful, ominous arcs. On other visits, we've stopped to watch bald eagles gliding on wind currents above the river.

From Lookout Point, Sentinel Rock measures empty sky in the dizzying space that separates it from the bluff.

Mississippi Palisades State Park gives witness to the slow wheel of geologic time. But when you catch it just right—when an eagle sweeps in close view beyond the bluff, or when a huffing train strains past two hundred feet directly beneath—time is also measured in captured moments.

## Upper Mississippi River National Wildlife and Fish Refuge

A few years ago, I found myself with a spare day on my hands in mid-October. When I arrived at the Spring Lake overlook at the Upper Mississippi River National Wildlife and Fish Refuge on a clear October morning, the ducks and geese splashing and honking away in the Mississippi backwater sounded curiously like kids at the city pool.[3] Summer was fading, but not, at this point, any too quickly.

I saw no long V-formations across the sky. No one was in a hurry to leave.

Ed Britton has been district manager for over twenty years at the Upper Mississippi River National Wildlife and Fish Refuge–Savanna District. "I love being on the river," he said, and from his perch of longevity he's been able to watch conservation projects birthed, nurtured, and come to fruition. And he's seen the river suffer environmental degradation as well.

The Upper Mississippi refuge stretches 261 miles from Wabasha, Minnesota, to Rock Island, Illinois, covering two hundred forty thousand acres of backwaters and islands. It is divided into four districts, of which the Savanna (Illinois) District constitutes the southernmost eighty miles.

The Savanna District typically sees tens of thousands each of canvasback, mallards, Canada geese, and other migrating waterfowl each fall and spring as they use the Mississippi River valley as an arterial flyway. The refuge provides rest havens and sanctuaries. Certain areas, like the thirty-six-hundred-acre Spring Lake, are off-limits to hunting for migratory birds.

In contrast to the upper three divisions, the Savanna District has a much wider flood plain and is not as hemmed in by river bluffs. Unlike the northern districts, it has gated levees, "not to keep the water out," said Britton, "but to let water in. The levee system is used to raise and lower backwater levels to best serve migratory waterfowl." Backwater levels are lowered in the summer to encourage plant growth, and then raised in the fall and winter for migration seasons so that geese, ducks, and other waterfowl can have a swim-up-and-dine experience. The levee backwaters are likewise managed for fish, shore birds, and waterfowl. Hundreds of bald eagles overwinter on the refuge.

The Savanna District also includes four thousand acres of upland sand prairies formed by ice dams during glacial periods. The resulting glacial lakes dropped enormous amounts of sand on the lake bottoms, which remained on today's uplands when the ice dams melted, the glacial lakes drained, and the Mississippi River assumed its modern valley. Some of the sand prairie is located in the Lost Mound Unit, that is, the former grounds of the Savanna Army Depot. Much of the Army Depot is now part of the refuge but largely off-limits to visitors due to soil contamination and areas of potential unexploded ordnance from its army legacy.

The sand prairie, largest in Illinois, continues south of the former Army Depot as well in uplands surrounding the refuge's visitors' center. Three miles of the paved Great River Trail wind through the sand prairie, offering bicyclists and hikers eye-level views of big bluestem and prairie compass plants. Pull-out spots provide river overlooks, shoreline hiking, and informational kiosks explaining the ecosystem of the prairie and its protection of rare species like the ornate box turtles that had already burrowed into the sand for the winter on this crisp October day.

Although I had other tasks to attend to back home, I finished my day at the refuge with a thirty-mile bike ride on the trail. The sky was a crisp blue devoid of summer haze, the day neither too warm nor cold. Ed Britton likes it here. The geese and ducks—for a while at least—were in no hurry to leave.

Why should I?

## Black Hawk State Historic Site (Saukenuk)

The trees were already barren in Dubuque several years ago when Dianne and I left for a late-autumn hike at the Black Hawk State Historic Site in Rock Island.[4] But here, just seventy-five miles south of home, the leaves were taking a last valiant hold. You knew it couldn't last.

The Black Hawk site occupies 208 acres along the Rock River near its confluence with the Mississippi. The 160-acre forest lies only minutes away from the historical location of Black Hawk's Sauk village, Saukenuk.

Saukenuk in 1832 was no small outpost of nomadic Native Americans, but a hundred-year-old established village of thirty-five hundred. The Sauk had migrated to the Upper Mississippi from the Great Lakes in the early 1700s. Their new village sat strategically near the confluence of the Rock and the Mississippi, offering river highways to the north, south, and northeast.

Saukenuk was laid out like a modern city, with parallel streets and intersecting alleys, a council house, and a public square. Outside the village, women tended eight hundred acres of corn, pumpkins, squash, and other vegetables. The men fished and hunted. Through the spring, summer, and autumn, villagers engaged in games and contests, and the young courted and slipped off to the nearby Rock Island to pick berries.

In late fall, the villagers dispersed in small bands into the Illinois prairies or crossed the Mississippi into Iowa to overwinter, hunting for meat and furs. In the spring, they returned to Saukenuk, and the cycle repeated.

Today, the Black Hawk Forest is noted as one of the least disturbed woodlands in Illinois with its thickly girded oak, maple, and hickory hardwoods. Here, an eighty-foot oak with a five-foot circumference may have actually seen the days of Saukenuk.

Almost thirty-five wildflower varieties bloom in the woods and reconstructed prairie. Native tallgrass big bluestem tower above eye level. Birders have documented nearly 175 species, especially in spring migration season.

In the bluffs above the river once lay two-thousand-year-old burial mounds of Hopewellian peoples who predated the Sauk. The mounds are no longer intact.

A pioneer cemetery on the edge of the grounds reminds you that Saukenuk would not last. Change was in the air.

The Sauk had not won the graces of the new American government, having sided with the British in the Revolutionary War and again in the War of 1812. The Americans built Fort Armstrong on Rock Island to keep an eye on things.

In 1804, a group of five Sauk trading in St. Louis were forced to sign a treaty that they themselves did not understand and had no authority to sign. The treaty ceded—at some undetermined future date—all land east of the Mississippi, including the village of Saukenuk, to the Americans. Since the treaty would not go into effect until settlers arrived, and since its signatories hadn't understood it anyway, it was unknown among the Sauk until they were informed in 1831 not to return to their village the following spring.

Chief Keokuk counseled the Sauk toward peace and acquiescence. Black Hawk, a prominent warrior of the tribe, disagreed, and with a band of fifteen hundred men, women, and children, including the elderly, set back across the river in April 1832. He'd been led to believe that other tribes would join with him in reclaiming the old Sauk lands, and that even the British would send supplies from Canada.

When no support materialized, Black Hawk tried to sue for peace, sending a team of five Sauk with a white flag, trailed by another five observers. When militiamen spotted the observers hiding in the woods, they mistook it for an ambush and attacked. A few Sauk escaped and returned to Black Hawk, who retaliated. The war was on, based on a botched surrender.

It was more of a chase than a war, really. Black Hawk quickly realized that he could not succeed without the help he'd expected, and so began to march his band up the Rock River to escape the militia and the army, which was always a day or so behind. On the move for several months, the Sauk began to starve, and eventually decided to head west to the Mississippi.

The whole affair ended horrifically nearly four months later at the Mississippi River when the US military caught up with the Sauk at a site between Prairie du Chien and La Crosse, WI. American soldiers shot from the shoreline and from a riverboat as Sauk women, children, and elderly tried swimming to safety across the Mississippi. Between the massacre, some earlier battles, and general starvation, only 150 of the original 1,500 Sauk survived at the end of the Black Hawk War.

But for now, on this late-autumn hike with my wife at the Black Hawk Forest, the canopy was ablaze in a sun-drenched golden brown. Driving home, however, we noted again the barren trees further north and realized that the old village of Saukenuk, too, would soon experience the fall.

## Pere Marquette State Park and Two Rivers National Wildlife Refuge

We were near Cahokia when we first found Pere Marquette State Park and the Two Rivers National Wildlife Refuge. The park is named after Fr. Jacques Marquette; he and Louis Joliet were the first Europeans to enter the Upper Mississippi. The refuge is another stop-over sanctuary for waterfowl using the river as their migration highway.

Marquette and Joliet, along with a crew of five voyageurs, entered the Upper Mississippi on June 17, 1763. They had ascended the Fox River from Green Bay in two canoes, portaged to the Wisconsin River where the two rivers flowed near one another, and then descended to its confluence with the Mississippi near present-day Prairie du Chien.

The voyage was full of new discoveries. Marquette sees "on the water a monster with the head of a tiger, a sharp nose like that of a wildcat, with whiskers and straight, erect ears," which (minus the ears!) seems to have been a large catfish.[5] He spies the "wild cattle [which are similar to] our domestic cattle. They are not longer, but are nearly as large again, and more Corpulent.... Under the neck they have a sort of large dewlap, which hangs down; and on the back is a rather high hump."[6] He is, of course, describing the creature for which the French had no name as yet: the American bison.

Marquette and Joliet continued down the Mississippi along the future shores of Iowa, Wisconsin, Illinois, and Missouri. Between the mouths of the Illinois and Missouri Rivers, a little south of today's park, Marquette grew fearful of the petroglyph he spied on the exposed bluffs on the Illinois shore. He was seeing the Piasa Bird:

> While skirting some rocks, which by their height and length inspired awe, we saw upon one of them two painted monsters which at first made us afraid, and upon which the boldest savages dare not long rest their eyes. They are as large as a calf; they have horns on their heads like those of a deer, a horrible look, red eyes, a beard like a tiger's, a face somewhat like a man's, a body covered with scales, and so long a tail that it winds all around the body, passing above the head and going back between the legs, ending in a fish's tail.[7]

The original Piasa Bird was destroyed by modern quarrying. Dianne and I spotted the Piasa reproduction along the twenty-mile Sam Vadalabene Bike Trail between Alton, IL, and Pere Marquette State Park. It looked menacing enough, though it seemed to me to promise havoc only to those who would ever again destroy the bluffs.

Sometimes physically separated from the road and sometimes merely a wide side lane, the Vadalabene Bike Trail parallels the Great River Road.

We had driven the road the night before and watched the full moon splay its reflection across the Mississippi like a great barge floodlight. On our bikes, in daylight, we stopped to watch the sun rippling across the waves. The Illinois bluffs near Alton, IL, undulate along the Mississippi shore rather than putting up a flat-fronted cliff face. Their yellow-white limestone edifices repeatedly curve toward the river and back away again like the swell of the river waves that cut them. The result is a series of castellated rock towers overlooking the river.

Marquette, Joliet, and the crew pressed onward until present-day Arkansas. Not wishing to proceed into Spanish territory, they turned back. On the return trip, Marquette and Joliet took the shortcut route up the Illinois River to return to Lake Michigan.

The eight-thousand-acre Pere Marquette Park begins near the mouth of the Illinois River, where the French explorers diverted from the Mississippi. The bike trail took us from the riverbank up into the park. We rode its inland swells and dips as the trail assailed the bluff and dropped down again. Eventually we left the bikes behind and climbed to the Eagle Roost Overlook for a view of Gilbert Lake, the Illinois River, and Swan Lake in succession.

On another day's drive we crossed the Illinois River on the Brussels Free Ferry. A late-September morning gave us the ferry to ourselves, lazing across the river on the five-minute shuttle before arriving at the ninety-two-hundred-acre Two Rivers National Wildlife Refuge nestled in the peninsula between the Illinois and Mississippi Rivers. It, too, is a major stopping point for migratory waterfowl in the fall and spring.

We took a short hike from the visitors' center. A dozen quail erupted from the edge of the trail as we passed, startling us with their gunshot departures. Monarch butterflies alighted on the wildflowers while big bluestem towered above our heads. Milkweed seed pods were beginning to open. Lotus flowers floated on the refuge's ponds. In the distance, waves of gulls floated over the rivers.

This was the northern edge of what is called the American Bottom, the fertile floodplain that runs for seventy miles where the Illinois and Missouri Rivers meet the Mississippi. Two Rivers National Wildlife Refuge this day was flush with wildlife but nearly empty of humans. But in an earlier age, the American Bottom was the seat of the largest pre-Columbian civilization and city in North America.

---

We'd inched our way down and back along the Illinois shoreline to these and other Illinois locations through four decades of marriage, returning to

Curved bluff walls line the east shore of the Mississippi River near Alton, IL.

favorite haunts across all four seasons with bikes, boats, and boots. Each was a quilt piece, an individual patch, a separate story stitched inexorably to the next.

Still missing was Cahokia, the prize civilization that had flourished and perished along the river in the pre-Columbian past.

## Cahokia Mound City

At long last, after many years of intentions and dashed plans, we finally arrived at the Cahokia Mounds State Historic Site near Collinsville, IL, in September 2021. The slim fallen leaves of a young walnut tree—first victims of the encroaching autumn—swirled in circles at our feet as we approached the visitors' center. We soon lined up inside with a group to tour the grounds under the tutelage of a young archaeology graduate named Matt.

Outside, the morning was already blasting warm for September, around eighty degrees Fahrenheit and pushing higher. We clustered in the building's shade while Matt set the context. Human habitation in the region, as throughout most of the Midwest, he said, had begun twelve thousand years ago as the glaciers to the north began their retreat. Cultures evolved over time from hunting and gathering to cultivation, adapting as the climate gradually grew warmer.

We ventured out onto the gravel path. The grounds were a blistering green, the grass short-clipped. The mounds, differently sized and shaped, seemed scattered. A long empty field stretched before us. In the distance I could see Monks Mound, the iconic, stair-stepped Cahokian structure I knew best prior to our visit.

The city known today as Cahokia, Matt told us, took hold around AD 800 and reached its zenith around 1050 as a renaissance took hold in the American Bottom. At a base level, the renaissance stemmed from an agricultural revolution focused on growing the "three sisters" of beans, squash, and corn. The development of an agricultural diet allowed for greater food storage, enabling large populations to live together. Freedom from the constant search for food in turn fueled the renaissance in architecture, artwork, leisure activities, and spirituality.

Artifacts unearthed at Cahokia place the city squarely in the Mississippian culture that flourished from Florida to Wisconsin at that time: the bird man tablet with its imagistic references to the upper, middle, and underworld in the forms of bird, man, and serpent; and pipes with animal and human effigies. Other ceramics whose patterns and styles associated Cahokians with faraway places suggested extensive trade, immigration, and cultural interaction.[8]

Cahokia wasn't an anomaly as a city of its time, but it was the most significant metropolis, perhaps the center, of the Mississippian culture.

The American Bottom positioned Cahokia for exponential growth. Annual flood silt kept croplands fertile.[9] Old Mississippi meanders and oxbows became lakes, sloughs, and marshes, suitable for fishing, clamming, and gathering edible cattails. Nearby streams offered direct access to the Mississippi, which in turn—along with the Illinois and Missouri Rivers—provided trade and migration routes eight hundred miles in every direction. These routes would bring copper from the Great Lakes, mica from the Appalachians, and seashells from the Gulf of Mexico.[10]

The city, Matt explained, was also near prairies to the north, which provided not only game but also thatch for the many homes' roofs, and to woodlands to the south and east for timber. The Cahokians felled many trees for their various needs, but spared pecans and walnuts for their fruits.

The city itself covered four thousand acres with at least 120 mounds. It ran three miles east to west, and more than two miles north-south. Its population, peaking between AD 1050 and 1200, may have reached twenty thousand, with some archaeologists putting it at thirty thousand.[11] Smaller villages surrounded Cahokia with twenty thousand additional people, like modern suburbs. But Cahokia was the cultural and civic capital.

Matt led us among Cahokia's most visible feature, its wealth of mounds, nearly seventy of which are still preserved at the Historic Site. They popped from a level field in the city center, round-topped, flat, or layered. Some mounds stood nearly as tall as adjacent trees. The Cahokians even artificially sculpted the level field from the mottled knolls and dips that naturally resulted from swirling waters and siltation during floods. And during the somewhat drier era in which the city flourished, annual floods typically didn't reach as far inland as Cahokia.[12]

The city boasted three types of mounds, all of which were laid out before us. Conical mounds were similar to those along the upper Mississippi, though taller at up to forty feet high. Many of these held burials.

The second type were ridgetops, oval-shaped mounds capped off with angular ridges like tent-tops or rooflines. A half dozen or so still existed at Cahokia. The ridgetops may have been added to conical mounds when their original function had been completed. Ridgetop Mound 72, for example, sits atop three separate conical burial mounds. The mounds, excavated prior to the 1990 Native American Graves Protection and Repatriation Act, contain nearly three hundred burials. These burials include a sacrificial mass grave with fifty-three women laid out two deep in two rows, another with four men shorn of their heads and hands, and a burial of an elite man and woman laid on a bed of twenty thousand marine shell beads and four hundred ceremonial arrow points, along with six attendants.[13]

But platform mounds particularly define and identify Cahokia and the Mississippian culture. Platform mounds resemble pyramids without their angular peaks. Temples, council houses, and the dwelling places of shamans and high chiefs typically adorned the flat platform at the top of these mounds.

Monks Mound is premier among the platform mounds at Cahokia. Matt pointed in its direction but wouldn't be taking us there as a group, due to the challenges of everyone climbing its 154 stairs to the top. Standing 100 feet tall on a base of 1,040 by 800 feet (nineteen acres), it is the largest Native American mound in North America, third largest in the world, and slightly larger than the Great Pyramid of Egypt and the Pyramid of the Sun

in Mexico. It was constructed by Cahokians carrying 22 million cubic feet of soil, sand, clay, and rock, basket by basket, from nearby floodplain "borrow pits,"[14] similar in this respect to all of the mounds at Cahokia. Some borrow pits were retained as ponds, but others were filled in with trash.

The long, open field we'd noticed running from the base of Monks Mound to the Twin Mounds where we now stood at the opposite end was the sixteen-hundred-foot Grand Plaza, a rectangular public gathering space. Cahokians likely came to the Grand Plaza for markets and festivals, and for religious, civic, and sporting events. Acoustics in the valley could carry a booming voice from the top of Monks Mound out over the Grand Plaza. Matt took our imaginations back to some ancient feast with the high chief and shaman addressing thousands of Cahokians from on high, the sun glittering off metal ornaments at the top of the mound.

The gravel path that Matt led us on ran partially along the route of the stockade, a two-mile log wall that enclosed the interior of the city. The massive structure would have required 15,000–20,000 tree trunks, all brought in from beyond the city's edge. The stockade's purpose isn't fully

Monks Mound at Cahokia, IL, is the largest pre-Columbian, human-built earthwork in North America. At the top of this platform mound, a large temple or dwelling place of a high priest or chieftain towered over a city of twenty thousand people around AD 1100.

understood. It may, of course, have served a defensive role, but there is little evidence of extensive regional warfare during the period. It appears, though, to have divided the city's hierarchy, with elites living within the enclosure and commoners on the outside in homes of pole and thatch construction, with areas for toolmaking, cooking, storage, and sleeping.[15]

The Mississippian culture spread well beyond Cahokia, most likely through a combination of emigration, trade, and cultural exchange. Wisconsin's Aztalan State Park has three platform mounds at a distance of three hundred miles north of Cahokia. Trempealeau, Wisconsin, nearly five hundred river miles to the north, likewise has three platform mounds overlooking the Mississippi, as well as artifacts directly linking the builders back to Cahokia. Mississippian influence even spread to the distant southeast.

Loras College professor of history Dr. Kristin Anderson Bricker depicts Cahokia's spreading influence as a cultural osmosis rather than geographic hegemony. These distant locations, she said, "are often referred to as 'colonial outposts,' but a big piece of Cahokia's influence was due to their trading networks. There may have been cultural elements brought back and forth by travelers. But I don't think you could argue that it was a territorial empire."[16]

Cahokia's collapse after AD 1300 remains a mystery. There are several likely culprits, perhaps in combination. The city may have "collapsed under its own weight," writes William Iseminger, as nearby forests and other resources were depleted.[17] Ill health may have resulted from water and air pollution and limited diets, all of which might have been mitigated in smaller communities. A cooling climate after 1250 may have led to shorter growing seasons, causing malnourishment.[18] And the cooling climate might have been accompanied by more frequent, destructive floods that reached the city proper.[19] Other locations might suddenly have looked more promising. Big game like bison, for example, were more common to the west. Perhaps Cahokia was beset by enemies, although no archaeological evidence points in that direction.

As Cahokians dispersed in different directions, they became or melded into historic-period tribes such as the Osage, Omaha, Kansa, and Missouri. Thirteen modern tribes claim Cahokian ancestry.

The physical remains of the metropolis nearly disappeared. As St. Louis emerged and grew, its settlers razed mound sites west of the Mississippi. Of forty mounds thought to have been located in the St. Louis area, thirty-nine were destroyed. Some Europeans thought the mounds to be natural features that needn't be preserved. At Cahokia itself, French priests built a chapel on the first terrace of Monks Mound in 1730.[20] French Trappist monks then took up residence on the grounds from 1809 to 1813, even growing crops on Monks Mound, hence giving it its modern name. (Cahokia itself takes its name from the unrelated Cahokia tribe who lived nearby in the late 1600s.)

Cahokia continued slipping toward erasure from the landscape. A house was built on the top of Monks Mound in 1831. By the 1860s, much of the former city was being plowed and farmed. Powell Mound was destroyed during 1930–1931 to fill in a lowland for tillage (the lowland quite possibly being the borrow pit from which the mound was built in the first place).[21] Subdivisions encroached on the ancient city. A modern four-lane highway slashes through the Cahokian grounds, passing at the foot of Monks Mound. Interstate construction plans in the 1960s would have included an interchange at the edge of the grounds until its location was shifted.

But a preservation movement was already underway. Beginning in the 1830s and with each new slice of mound slated for destruction, archaeologists were establishing the extent and radiance of the city. The state of Illinois finally purchased 144 acres in 1925. In the ensuing decades, the state and the Cahokia Museum Society acquired fifty-five additional parcels, bringing the Cahokia Mounds State Historic Site to its present size of twenty-two hundred acres. Cahokia became a US National Historic Landmark in 1965 and a UNESCO World Heritage Site in 1982, further ensuring its preservation.[22]

Here it was, laid out before us, this place we had heard and read about and thought we understood. But "you understand a subject differently when you experience its physical space," Anderson-Bricker said.[23] When she and her artist-husband visited Cahokia, he created panoramic views from his own photographs, which she uses in her teaching.

Being physically present in the space likewise conveys the complexities of a civilization. It helps modern students understand that "there are different cultures, not better or worse ones," Anderson-Bricker believes. And it creates a sense of wonderment and awe, and lessens our certainty that we know and understand everything about a culture. "Most of our knowledge is through the lens of White archaeologists and anthropologists," she adds. Recognizing our own cultural slant demands a certain humility and awareness that "we may have this wrong."[24]

Later, Dianne and I took our own walk through the Cahokian grounds. Over here had sat the huts of the common people. Over there ran the wooden stockade. Before us lay the Grand Plaza. We hiked a trail winding in and out of the woods where we found more mounds in a remote clearing. Deer materialized at the edge of the forest, perhaps relieved that they'd arrived seven hundred years too late to be hunted. We walked across the road to the Cahokia Woodhenge, a reconstructed circle of forty-eight red cedar posts configured to mark the equinoxes, solstices, and other celestial events. I stood at the center post, aligning it with the perimeter post in the shadow of Monks Mound, where the fall equinox sun would rise. I half-jokingly suggested returning at sunrise, as we were just two days short of the equinox. Dianne diplomatically put the notion on the back burner rather than rejecting it outright. We needed, after all, to get back on the road home the next morning.

But we'd finally found the center quilt patch, Cahokia, the city that had thrived near the Mississippi River in the American Bottom nine hundred years ago, was abandoned, and had nearly disappeared from the landscape. The big patch was faded, but was still the prize.

Driving home to Dubuque by way of the four-hundred-mile Illinois shoreline route along the Mississippi River was a bit like running one's fingers across the stitching of seemingly disparate memories, peoples, cultures, and times. We stayed overnight in Alton, founded in 1818 at the site of a Mississippi River ferry service. We took today's ferry across the Illinois River, drove through the Two Rivers National Wildlife Refuge, and got ourselves as far and deep into a mud-and-rock road as our Honda CR-V would take us. We stayed overnight again in Quincy, founded in 1822 and famed as a riverboat stop. A bike ride through and around town took us past nineteenth-century mansions and Hopewellian mounds.

We stopped in Rock Island to revisit the forest where Black Hawk's village of Saukenuk once stood. We went to Albany, a town of nine hundred where thirty-eight Hopewellian mounds remain on the bluffs above the river. We drove through Savanna, founded in 1828, likewise as a riverboat town. We edged past the Upper Mississippi River National Wildlife and Fish Refuge–Savanna District, where we'd been bicycling the week before. And past the former grounds of the Savanna Army Depot, where ammunition had been built, tested, and stored at various times between 1916 and 2000, and which now cradled a wildlife preserve amid contaminated grounds.

We drove through Galena, an 1826-founded lead-mining town. We drove through East Dubuque, once called Dunleith, our highway passing beneath four large Hopewellian mounds on the river bluff at Gramercy Park.

And we crossed the Mississippi River again on the Julien Dubuque Bridge, arriving back home in Dubuque, Iowa's oldest city.

They were all places we'd visited in separate trips over several decades. Now they felt whole, mile blending into mile, generation blending into generation, although with a line of stitching ripped at the forced displacement of indigenous peoples.

But as far back as the mind can fathom, there has been the Mississippi River stitched alongside them.

We'd saved climbing Monks Mound till the end of our visit to Cahokia. The 154-step stairway ascends the mound in two segments divided by a level terrace halfway up.

At the platform's top, the view stretches in all directions. Etched stone markers suggest what would have lain within sight in AD 1100. To the north lay the stream that linked the city to the Mississippi River. To the east were dwelling huts, the stockade, and more mounds. To the west, the Cahokia Woodhenge and the Mississippi River. To the south, the Grand Plaza stretched out before the mound the length of a football field. Thousands of Cahokians might have gathered there to catch a glimpse of the high chief or shaman on a festival day.

Today, of course, the St. Louis Gateway Arch is likewise visible on the western horizon, completed in 1965 to commemorate American westward expansion.

Whether a shaman with a vision of the future might have seen or sensed it back then, we'll never know.

# 9

# North Dakota

*The Rendezvous Region: The Meeting Place*

Interstate 79 races north toward the Canadian border at seventy-five miles an hour near Pembina [PEM-bi- na], North Dakota. The road and its setbacks seem wide here, the plains even wider. Occasionally, the Red River of the North creeps within a mile and a quarter of the road, and that faint line of trees to the east might mark the riverbank. We haven't seen a car or truck in either direction for the past ten minutes. Within a 360-degree sweeping view we might see a single farmhouse, or maybe not, although the numerous single-file tree lines in the fields suggest they may have once sheltered smaller farmsteads.

Yet here we are in the Rendezvous Region of northeast North Dakota, where Euro-American fur traders, Native Americans, and north-flowing rivers all came together—sometimes productively, sometimes clashing. The meeting place.

---

Dianne and I entered North Dakota from Minnesota, crossing the Red River into Fargo, the state's largest city at 125,000 people. Fargo was too much of a meeting place for us, as evidenced from the bustling downtown and interstate traffic. We continued up the interstate to Grand Forks (population fifty-six thousand, third largest in North Dakota) for an overnight stay after nine hours of driving, and for a reasonable embarkment to the north come morning.

Grand Forks, too, was a meeting place, that of the Red Lake River homing in from Minnesota and the Red River of the North that marks the boundary between the two states. This confluence of rivers has been both fortuitous to Grand Forks—giving the city its name and its founding purpose—and occasionally disastrous.

The rivers' confluence had made the Grand Forks area an important Native American trade center for thousands of years prior to the arrival of Europeans. By the mid-1700s, French explorers, trappers, and traders had dubbed the region *Les Grandes Fourches* as they, too, traded with the indigenous population and moved furs north to the Hudson Bay and east to the Great Lakes, to the Atlantic colonies, and on to Europe.

The city didn't take root, though, until the American steamboat era. For reasons that only a northern midwesterner could understand, when steamboat captain Alexander Griggs and his crew found their boat trapped overwinter in the ice at the confluence in 1870, he decided that *this* would be a good place to establish a town. Why ever not?

The Red River retains its name after the confluence, and flows north to the Canadian border and into Lake Winnipeg. From there its waters inch toward Hudson Bay, out to the Labrador Sea, and into the North Atlantic Ocean. North-flowing rivers aren't rare across the globe (think of the Nile or the Rhine), but in the United States the largest watershed sends most rivers in a southeasterly or southwesterly course to join the Mississippi River heading south to the Gulf of Mexico. A thin continental divide in the upper Midwest sends a small number of rivers toward Hudson Bay or to the Great Lakes and St. Lawrence River.

While all upper Midwest rivers are prone to spring flooding as winter's snow and ice let loose, northward flows like the Red River face a unique spring flooding phenomenon. South-flowing rivers melt from south (downriver) to north. So as their northern upriver reaches finally begin to melt, there is someplace for the meltwater to go, namely downstream. They may flood briefly, their banks unable to handle the flow, but there is an escape route for the high waters within a few days. In a north-flowing river, the southern (upstream) portion melts first. The meltwater flows north until it meets solid ice, which acts like a dam. The river backs up, and its spring flood lasts longer, until the northern ice sufficiently melts to allow smooth drainage.

Grand Forks has seen many epic floods. The Red River's most devastating floods have recurred with some regularity, the worst ones prior to the current century having occurred in 1882, 1897, 1950, 1996, and 1997. Twenty-first-century flooding has accelerated the trend. The past two decades have seen an increase in high-water marks, although now contained in Grand Forks by its flood wall and greenway. Flood-stage crests in 2006, 2009, 2010, 2011, and 2019 include the third-, fourth-, and eighth-highest levels on record.

The 1997 flood was the worst on record at Grand Forks with a fifty-three-foot crest. Nearly the entire city's population had to be evacuated, the largest evacuation of US citizens from the Civil War until Hurricane Katrina. When a simultaneous fire broke out in a downtown business building, it quickly spread to ten additional structures. Firefighters struggled to contain it, as their equipment was ironically hindered by the high water.

Over fifteen hundred homes and businesses were destroyed in the 1997 flood. Another 850 surviving properties participated in a buy-out that allowed for creation of an eight-mile, twenty-two-hundred-acre flood plain greenway backed up—at some distance—by nine hundred feet of floodwall designed to protect against a sixty-three-foot flood. The greenway was transformed into a park-like space along the river with twenty miles of walking and bicycling trails.[1] Due to the floodwall and green space, twenty-first-century flooding hasn't had nearly the same economic and human impact.

Upon our arrival in Grand Forks on that first late afternoon after nine hours of driving, exhaustion warred with our bodies' need to stretch. The bikes on the car rack won out. The trail winds gracefully through the greenway, wide enough for cyclists, walkers, and roller bladers to mind each other's spaces. Here it caresses the floodwall, which is surprisingly aesthetic with its faux limestone façade. And there the trail dips down to the riverside or crosses one of the three bridges (two of them nonvehicular) into the East Grand Forks side of the greenway.

The Red River on this late-June day was running swiftly after recent rains, but nowhere near flooding. Still, it skipped and leaped at points, rushed under the bridges, and tickled at low-hanging branches and shoreline grasses. This was a river with business to attend to.

The seven-story observation tower at the Pembina Museum gave us a ten-mile view of the Red River valley in every direction, including a glimpse well past the Canadian border just two miles to the north. To the southwest the view takes in the Pembina River that feeds into the Red River.

The view is nowhere hemmed in by hills because there virtually are none. The Red River valley stretches some 30–40 miles inland here near the Canadian border, which is partly why its frequent floods are so devastating. It can and occasionally does spread out over much of its flood plain. But the wide valley was not carved by the flooding river. As the glaciers were melting from North Dakota twelve thousand years ago, the Red River valley lay at the bottom of Glacial Lake Agassiz for about four thousand years. Lake sediment slowly filled the lowlands, leaving the land with a flat, fertile prairie soil when the glacial lake finally drained into Lake Winnipeg and Hudson Bay. The soil is sandy and silty, with a hint of fine clay.[2] We could feel this finely textured soil beneath our feet in the city park, where the grass is thinned by a fairly dry climate that receives only twenty-one inches of precipitation a year. The crops were a little foreign to us from our part of the Midwest where corn is king. Here, the cultivations of choice include wheat, sugar beets, sunflowers, and potatoes.

Just five hundred residents live in the town of Pembina, and sixty-seven hundred in Pembina County. The town was preparing for a major blow, though, as Motor Coach Industries had recently announced it would be closing its Pembina factory at the end of 2022, to a loss of two hundred manufacturing jobs.[3] As we left the town park, we saw two or three finished touring buses pulling out from the factory doors en route to their next destination.

Although never heavily populated due to the harsh climate, this corner of North Dakota was known to an earlier era as the Rendezvous Region. Prior to European contact, Ojibwe, Dakota, Assiniboine, Cree, and Cheyenne tribes moved in and out of the Red River valley on a seasonal basis. But the arrival of French and British—and later Canadian and American—fur trappers and traders turned the Red River valley into an annual meeting place for trade and transportation of pelts north and east to markets and seaports. Later, beaver pelts were supplanted by bison hides. From 1800 through the late 1870s, oxcarts carrying the hides dispersed from Pembina after the annual rendezvous. The carts were known for their creaking wheels that were left ungreased so as not to muck up from the prairie dust. The oxcarts could also be converted into rafts, as necessary, for river crossings. Today their ruts still occasionally haunt the landscape.

Pembina was a cultural meeting place as well. French traders in particular married Native American women, resulting in a large Métis population whose culture drew from European and indigenous traditions but was also distinct from both. The Métis were known for their colorful dress and lively fiddle-based music and dance. The Pembina area's Métis population grew from five hundred in 1821 to twelve thousand by 1870. The Métis controlled the oxcart transportation of goods.[4]

Leaving the town of Pembina, we stopped briefly at the reconstructed trading post and cabin home of Antoine-Blanc Gingras, a Métis leader of the town of St. Joseph, where fifteen hundred Métis once lived amid the fur and hide trade in the 1840s. Today the view from the trading post and cabin—a few miles removed from St. Joseph—looks more like it probably did presettlement, with tall grasses waving in the wind all the way out to the horizon where a line of trees marks the Pembina River. A small indentation in the landscape before us is an oxbow, an abandoned loop of the river.

Gingras had supported a Métis national independence movement in the Rendezvous Region in recognition of their unique culture. But the movement failed as the beaver and bison disappeared, and along with it the need for oxcart transportation. The Métis began to disperse. Incoming Icelandic immigrants changed the town's name from St. Joseph to Walhalla in 1877, further erasing the Catholic Métis from the landscape.

The 1851 trading post of Gingras's business associate, Norman Kittson, likewise survives in Walhalla as North Dakota's oldest intact building. We found it in a corner of town amid a small but handsome oak savanna. The cabin's slumped, hewn-log siding and sagging, wooden-shake roof barely clung to the present age. We inhaled the old timber smell for a while. And then we headed for the hills.

The Pembina Escarpment puts a quick end to the Red River valley and the northeast flatlands. Outside of Walhalla, Highway 32 crosses the Pembina River and begins a long, curving ascent three hundred feet above the valley floor. While the top of the escarpment was likewise overlain by glaciers, its eastern slope was sharply gouged by the ice along a line of hardened rock. The escarpment slope later became the western shore of Glacial Lake Agassiz as the ice melted back. From the top of the escarpment, the Tetrault State Forest, covering over twelve thousand acres, looks down over the Pembina River winding and cutting through the highlands en route to the Red River valley. The Pembina's macaroni-shaped cutoffs through the gorge had been recharged from the recent rains, creating several oxbow lakes.

The escarpment's uplands undulate till they drop sharply into the gorge. At the Pembina Gorge State Recreation Area, we parked in a remote, dusty parking lot and hiked first through a rolling prairie under a sharp, blue sky. Small puffball clouds tried to punctuate the blue, but were no match for the sky that grows extrawide out here. Glacial erratic boulders scattered and nearly buried in the prairie soil reminded us that they'd been carried there and dropped by the long-ago ice.

Then we started on the trail that wound and snaked down through the valley and on toward the Pembina. The trail had been "helped" somewhere along the line by a bulldozer, so at its open faces we were hoping to win the paleontology lottery of spotting a sea creature fossil, perhaps a mosasaur, a 30-foot-long reptile with flippers that swam in the ocean here 70 million years ago. No such luck, but we definitely felt the descent through the ages. It was a warm day, in the low 90s, and we knew that each twist and curve downward along the trail would have to be paid back, with interest, on the climb back up. But curiosity kept leading us down one more bend.

We didn't hike all the way down to the valley bottom, but later we drove down a dusty gravel road that crossed the Pembina River at the historic 1905 Brick Mine Bridge. The old plank bridge spans the river with trusses the reddish-orange color of the bricks once mined from the clay here. The Pembina was lively today from the recent rainfall, and, catching the perfect angle through the crisscrossing trusses, we watched the sunlight play on its ripples. Behind us in the distance lay sharp cliffs the Pembina had once carved. Were we to go there, we'd no doubt find the fossil of a fifty-foot plesiosaur whose small tails and long flippers would have made them appear to fly through the seawater. But we had other places to be.

---

We set up our tent in Icelandic State Park after exploring the valley, the escarpment, and the gorge. A relatively new park, its initial property was donated to the state in 1963 by G. B. Gunlogson, who had been born to an immigrant family in 1887.

At the center of the nine-hundred-acre park sits Lake Renwick, a dammed impoundment of the Tongue River. Dianne knows that if there is water, I need to swish my feet in it at the very least, and if time and circumstances permit, my shins, my legs, waist, chest, and head. I have a high tolerance for cold, too, as I had discovered a few years earlier swimming in Lake Superior while Dianne lay on a blanket offshore, contentedly reading a book. These North Dakota waters were my siren call, too, and I gradually eased myself into the lake, acclimating inch by inch. The final baptism, though, requires a headlong plunge, and at last, I worked up the courage to take a dive.

Maybe I'm not as young as I used to be. Maybe my body fat, such as it is, isn't as insulating as it once was. After another dive or two I'd outlasted my compulsion and my tolerance and waded back out of the lake. "I think

The Pembina River has carved a gorge through the uplands leading to the Red River of the North in northeast North Dakota. Known during the 1800s as the Rendezvous Region, this was an annual meeting place of Euro-American fur traders, Native Americans, and Métis peoples to buy, sell, and transport furs to Europe and the eastern US seaboard.

I know why Icelanders settled here," I told Dianne, who looked at me as if I'd just proclaimed that the sky is blue.

The park lies adjacent to the two-hundred-acre Gunlogson State Nature Preserve, a woodland, prairie, and fen we passed while bicycling the paved Cavlandic Trail leading into the city of Cavalier.

We were on a mission in Cavalier. Our brother-in-law Dan from California had lived here briefly while growing up, just a few years, while his dad

had served in the military. We found the Cavalier Public School—home of the Tornadoes—and snapped a photo and sent it to him. Dan was touched by the gesture. He hadn't been there for forty-some years and never expected to see the place again.

The past is important here. Next to the Icelandic State Park Visitors' Center lies the Pioneer Heritage Center. It tells the story of the Gunlogsons, who departed from Iceland in 1876, tried their luck first in Gimli,

Manitoba, and then re-emigrated in 1880 with many of their countrymen to this locale. By the mid-1880s they'd replaced their original log cabin with a wood-framed house that G. B. Gunlogson lived in until his death in 1983. The still-standing home is on the National Register of Historic Places. The site includes a 1922 barn and a Lutheran church that once served the Icelandic immigrants.

Across the road lay another heritage site, the Pembina County Historical Museum. Dianne, an Iowa farm girl, traced the evolution of farm machinery, eventually finding the tractors, planters, and balers from her own childhood. I took a little more interest in the antique cars, including Gunlogson's own 1925 Case Model X Suburban Coupe. An evolution of fire engines resided in the building as well, watched over by two creepy mannequins in the back corner.

We took refuge in the 1882 Great Northern Railway Depot when a sudden rain deluge poured down. Dianne studied route maps while I rifled through purchase orders from the 1960s to see who had brought shipments of lumber, steel, and roofing into town. The rain wouldn't abate, though, so we finally took a run for it across the grounds to our car.

We were headed, or so we thought, to a place where the rain wouldn't matter.

The sky in North Dakota appears expansive, as if the dome above were taller, the horizons farther removed, than in the ordinary places. But when it's gray and raining, it closes in. The clouds seem barely above one's head. (I haven't been to North Dakota in winter, but I suspect that in a blizzard the sky disappears entirely.)

We drove eleven miles west of Icelandic State Park in the pouring rain to tour another feature of the North Dakota landscape, the decommissioned RSL #3 Sprint Missile site that was once part of the Stanley R. Mickelsen Safeguard Complex where Dan's father had been stationed as an engineer. The rain was slowing, but we still dashed from the car to the squat, concrete guardhouse, paid our twelve-dollar fee, and slid through the rotating iron bar gate that once served as a security clearance. The tour would take us below ground to the Operations Bunker, beyond the reach of today's gray skies and pelting rain.

The Mickelsen Complex was a six-site antiballistic missile unit that included four separate launch sites, of which RSL #3 was one. The other sites included a massive radar facility and general command center. At the height of the Cold War, the Department of Defense selected northern North Dakota as ground zero in case the Soviets ever launched ICBMs (Intercontinental Ballistic Missiles) in a nuclear attack. American Spartan missiles would, in turn, be launched from the Mickelsen Complex to destroy the ICBMs in flight. If the Spartans missed or the Soviet missiles were detected too late, Sprint missiles would launch from the RSL #3 site in a last-ditch attempt to defang the attack before it reached major metropolitan areas or our own offensive missiles located elsewhere in North Dakota.

A lesser destruction of human life on the North Dakota landscape was apparently acceptable.

Our tour guide and the decommissioned RSL #3 site's owner Mel Sann, a military veteran himself with a deep knowledge of the missile defense system, nevertheless called it "a train wreck waiting to happen."

Mel opened the steel doors and led us down a sloping seventy-five-foot tunnel to the underground bunker. With its massive concrete fortification, blast doors, and suspension-surrounded rooms, the bunker was designed to withstand a "10-megaton near miss one mile away and still function."[5] Inside were the gutted remains of the control room, underground living quarters, and decontamination showers. Wires and cables hung from the ceiling as if the departure had been a hasty eviction. Rust and flaking paint pocked the metal surfaces.

Bunkered against nuclear attack, though, the aging facility seemed challenged by rain and groundwater. Seepage sweated from the walls in the fifty-degree subsurface chill while multiple dehumidifiers hummed.

When we resurfaced, the rain had ended, so Mel took us out to the launch pad where sixteen small, domed lids in a four-by-four grid topped the now-empty underground Sprint missile silos. The long-gone missiles themselves had not been large, each measuring about twenty-seven feet long and only four and a half feet wide.

After showing us the launch pad guardhouse and the old security system rimming the chain-link fence, Mel led us back to our car. In the compound's grassy interior, he'd brought in sheep to keep the grass shorn, a herd of meekness in the midst of international tension.

Congress authorized the Mickelsen Complex in 1969, and construction began in 1970 at a cost of $6 billion. About thirty-two hundred civilians were involved in the construction process, and five hundred military personnel were stationed there, including Dan's dad.[6]

The local communities experienced both growth and burdens as construction ramped up, especially in the towns of Cavalier and Langdon. Langdon's population nearly doubled over a three-year period.[7] The influx of civilian workers and military personnel led to construction of seventy-two new family homes and 270 rental units. Secondary employment and new businesses serving the arrivals blossomed as well. But with this came municipal and community costs. Local water and sewage upgrades cost $1.3 million, and an "open dump" had to be replaced with a sanitary landfill. Roads needed improvement and schools required more space.[8]

The influx of residents and developed properties increased the tax base, but not enough to offset the demands. Since the Mickelsen Complex and its military housing were federally owned, they paid no local property taxes.

The Mickelsen Complex achieved full functionality on October 1, 1975.

The very next day, on October 2, the US Congress voted to *close* the facility as a result of nuclear arms treaties with the Soviet Union and amid concerns about the system's functionality. Senate critics doubted whether the system could reliably protect American cities, and called it "a system in search of a mission."[9]

The closing took about a year. The populations of Cavalier and Langdon dropped by 43 percent and 45 percent, respectively. School enrollments fell by 50 percent.[10]

The mayor of Langdon, John MacFarlane, complained bitterly to the *New York Times*: "We didn't ask them to come, but they did, and we did our best to accommodate them, and now they're going to leave."[11] The complex, they learned, had largely been a bargaining chip to bring the Soviets to the table. One of the complex's former units remains intact, the Cavalier Air Force—now Space Force—Station.

The region's loss of confidence in the federal government was severe. "There ain't nothin' gonna happen that'll straighten that out," MacFarlane continued in the *Times*. "That's gone."[12]

If the Rendezvous Region had once been a meeting place, this had been a colossal miss.

It began to rain again as we left the RSL #3 site. Not wanting to cook outside in a downpour, Dianne and I headed back to Cavalier and had dinner at the Cedar Inn Restaurant. Cavalier seems to have reached a new equilibrium these days. Its streets are well paved, the houses handsome and well kept. There is a new community pool in town and a new bike-share setup on Main Street. Unemployment in the county in June 2022 was 2.2 percent, well below the national average.[13]

~

After packing up from Icelandic Park, we drove into Langdon and took some photos of Dan's *other* elementary school from when he'd lived in the area. A replica full-scale Sprint missile sits in the park across the street. Asking directions, we struck up a conversation with two women out for a walk, both a bit older than ourselves. One of them had lived in Langdon all her life. She would have been a young adult when all of this transpired. The conversation was freewheeling: grandkids and schools and the summer weather. Then we left Langdon behind.

A stiff northwest wind was hauling a cold front behind it as we approached Grahams Island on Devils Lake on the southwest corner of the Rendezvous Region. A 2.5-mile causeway leads to the island, slicing the 330-square-mile lake (North Dakota's largest natural lake) roughly in half. On the west side of the causeway, four-foot waves that had been whipped up across the lake crashed onto the road base riprap, nearly spraying the highway. On the east side, the lake got a new reprieve, with mere ripples tickling the hulls of the fishing boats amassed there.

We set up camp in the park, wrestling the nylon tent into place against the twenty-seven-mile-an-hour winds and anchoring it with our sleeping bags and duffels. With the tent reasonably secure, we walked down to the lake where fishing boats were returning from a tournament, then swung inland to a woods and prairie hiking trail. Dark, menacing cumulus clouds mushroomed from the prairie horizon, and the wind-blown grasses swept along in waves. We hurried back to the campsite.

We usually like a more natural setting for our tent, but there was only one spot left in the campground when we made reservations, so ours was the only tent among the steel-sided RVs and trailers. After sparring with the cookstove in the wind, I returned to the car to write up some notes and Dianne retired into the tent to do some reading. The wind continued buffeting the tent. It genuflected, rose up, and pitched again and again. After a while we inspected the tie-downs that held the tent fly in place and found that they were beginning to fray.

Thus began the evening's entertainment for the RV crowd. As they watched from their sealed boxes, we crept through the wind to each of the tie-downs and duct-taped the straining fabric. Convinced that we had solved the evening's problems, we returned to our stations, me to the Honda and Dianne to the tent where the relentless wind flapped the nylon around her. Soon enough, Dianne opened the car door and proposed what would become the entertainment's second act, as we emptied the back of the car, dropped the back seats, moved our sleeping bags inside the vehicle, and placed all of the rest of our gear inside the tent.

There we spent the night, with just barely enough room in the car to stretch out (though not roll over). We'd silenced the wind.

The windows steamed over during the night as the outside temperature dropped. We'd been hiking under ninety-three-degree-Fahrenheit skies a few afternoons ago, and by morning we opened the car doors to a fifty-nine-degree-Fahrenheit sunrise. The tent had held through the night, and all our gear was still in place. But rather than mess with a campsite breakfast, we loaded up, dropped and folded the now-relaxed tent, and drove back across the lake to the town of Devils Lake and ate breakfast at a diner where the only wind came from a table of old men talking politics.

With a wave in their direction, Dianne said, "We need to go," and we hurried our breakfast along.

A second set of causeways farther east on the lake leads south and out of town, past the Spirit Lake Casino, and then curves westward toward White Horse Hill National Game Preserve. Along the way lie lands flooded by the decades-long increasing size of Devils Lake.

Devils Lake's natural basin is an endorheic, or closed, drainage area covering four thousand square miles in northeastern North Dakota. The basin is home to several lakes, the largest of which is Devils Lake. The basin has no natural drainage other than evaporation, and is recharged by rain, snowmelt, and natural springs. Ideally, evaporation and recharge would stay in rough equilibrium, but the reality is that the lake rises and falls with cyclical drought and periods of heavy rain. In historical times, the lake dropped precipitously in the drought-prone 1930s, bottoming out in 1940 at fourteen hundred feet above sea level. At that level the lake had severely receded beaches, and fishing was imperiled by the lake's increasing salinity. Then came a gradual rise through the 1990s, and then a quickly accelerating rise from the '90s to the present when it topped out in 2011 at 1454 feet. By 2007, the east end of Devils Lake had reached across six miles of low-lying farmland and valleys to combine with Stump Lake. Lake levels have fallen slightly since 2011.[14]

The current high-water levels are a fisherman's delight and a farmer's nightmare. Sharlene Breakey, a magazine journalist who grew up near Devils Lake, writes in *Modern Farmer* that backwater lands that used to dry up by midsummer in her teenage years are now under twenty to thirty feet of water. One farmer lost twenty-five hundred acres to the lake, and then regained eight hundred when the water level receded slightly. At least eight campgrounds are perched lakeside on former farm fields. Half-submerged barns haunt the water. A Google Maps satellite view shows Old Highway 281 largely underwater on the lake's west side. A new Highway 281 was built on higher ground.[15]

Devils Lake has dramatically risen and fallen all throughout its postglacial life. The drainage basin—covering about one-third of the state's northeast quadrant—was scooped out by the advancing ice. When the glacier began melting twelve thousand years ago, Ancient Lake Minnewaukan filled the entire basin as meltwater poured into the scooped-out hole.[16] Analysis of bank and lake bottom shows periodic swings from high to low volume every couple of hundred years.[17] Whether the current high-water period is part of that pattern or related to climate change is as yet unknown, but the rapidity of the increase has alarmed climatologists. The bathtub is filling more quickly than ever before.

---

We let Devils Lake's waves crash on the causeway as we pulled in at the White Horse Hill National Game Preserve.

The volunteer couple at the visitors' center told us where to look for the bison herd, except for one old bull who liked to wander off by himself and "rake you over with his stink eye" if you came upon him. So naturally that became our mission. Never mind the elk, the prairie dogs, the wood ducks, and the rest of the bison herd. We wanted to find Old Stink Eye.

The seventeen-hundred-acre White Horse Hill National Game Preserve in northeast North Dakota dates back to 1904 when it was originally designated a national park. In 1914, it was reclassified as a game preserve for nesting and migratory birds. In 1917, bison and elk were introduced to the refuge as their numbers had plummeted on the Plains. In the 1970s, prairie dogs joined the general melee.

Colleen Graue, visitor services manager for White Horse Hill, said that despite the preserve's modest size, it harbors three distinct habitats. Wetlands at the lake's perimeter are ideal for migratory birds. Elk love the woodlands. Bison and prairie dogs are partial to the prairie and savanna.[18] The preserve is home to over two hundred nesting and migratory bird species among all three habitats, but only a few hawks were braving the wind this day.

Hiking near the visitors' center along the fence of the bison/elk enclosure, we saw none of the refuge's dozen elk, but noted their handiwork in creating a cleanly cropped oak savanna in their recent munch-through of the perimeter woods. Woodlands are rare to North Dakota, where only 2 percent of the land is forested. But White Horse Hill's location on the southeastern edge of Devils Lake helped the woods get established in

North Dakota's White Horse Hill National Game Preserve is home to bison, elk, prairie dogs, and a wide variety of waterfowl and other birds. Its hilly terrain sits at the southern edge of North Dakota's largest body of water, the 330-square-mile Devils Lake.

presettlement times, Graue explained. Prairie fires that elsewhere swept across the landscape and kept tree growth at bay were snuffed out here by the big lake, allowing the forest to take hold on its southern flank.

The woodland trail from the visitors' center kept us on a ridge above the lake. This was a meeting place of forest species, blending typical midwestern oaks and ashes with more northerly basswood and aspen.

When we emerged from the woods onto a prairie bluff, the trail's descent toward the lake was blocked off. An old trail that once rimmed the lake was

now underwater, and the field beyond it was flooded. Dead, drowned trees lined the old shoreline and the hiking path. The lake's high waters have impacted the preserve in other ways as well. The visitors' center is new since 2006, necessitated by the rising lake waters, said Graue. The lakeside road to the old center is now underwater.

But the flooded lowlands have been a boon to the preserve's ducks, geese, pelicans, kingfishers, swans, water snakes, frogs, and other wetland species.

The preserve's 4.5-mile auto-tour loop within the bison/elk enclosure is a favorite for visitors. Here, beyond the wetlands, the prairie rolls across the valley floor. The grasslands are home to at least two prairie dog towns. When we stopped at the edge of their burrowed city, they seemed as curious about us as we about them. Perched on hind legs while chewing stalks of grass, they sized us up, wondering perhaps whether our Honda was a big white bison. Finally, though, the small engineers nosedived back into their burrows to return to the work of interconnecting the holes with tunnels.

In another section of prairie dog town, a dozen or so of the twenty-ish bison milled about the burrows, tugging at the sod and looking bored. A few calves among them pondered the prairie dogs with curiosity and amusement.

The bison herd is small, but the herd is prized throughout the federal Fish and Wildlife Refuge system for its genetic purity, Graue told us. And because the acreage is limited, the herd occasionally has to be culled. Bison from White Horse Hill have been sent to other national wildlife refuges. At other times, excess bison are slaughtered and the meat distributed at the nearby Spirit Lake Indian Reservation.

Near the edge of the perimeter hills, the prairie is dotted with hardwoods, mostly oaks, creating a prairie savanna. We climbed the 185-step stairway to the top of the preserve's namesake, White Horse Hill, overlooking the prairie's valley and the surrounding upland forest. A twenty-mile-an-hour wind raked the hilltop grasses in wavelike pulses.

White Horse Hill took its present name in 2019, the result of a long campaign by the local Spirit Lake Tribe of the Dakota. For a century, it bore the name of Sully's Hill, referencing Army General Alfred Sully, who had led a number of bloody campaigns against the Dakota Nation in the 1860s, including an 1863 massacre of four hundred Yanktonais, Isanti (Santee), and Hunkpapa in Whitestone, North Dakota.[19] Disturbed by the honor given to the massacring general within eyesight of the reservation, the elders campaigned to have the preserve's name changed. They put forth the name White Horse Hill, or *Sunka Wakan Ska Paha*, to commemorate a wild white stallion that in earlier times was known to come down from the hill to drink at the lake.[20]

We didn't see a white stallion. In fact, we'd nearly given up on seeing elk until we finally heard one through our open car windows, and then spotted him, slowly mashing his way through the forest, stripping and munching leaves as if they were lettuce.

Back at the prairie dog town, we had also looked for Old Stink Eye among the bison, but couldn't seem to find him. So we gave the auto-tour loop a second lap. We found him, finally, not far from the visitors' center, his dark, humped shape off by his sullen self in a grassy clearing.

He seemed to hold within that eye the preserve, the lake, the tribal history, all permanence and change.

Don't mess with me, he seemed to say. So we drove on.

---

Around the bend from White Horse Hill and across yet another causeway on the southern flank of Devils Lake lay the town of Fort Totten within the Spirit Lake Reservation of the Dakota Nation. On the edge of town sits the Fort Totten State Historic Site. The fort had served as a major military outpost from 1867 to 1873 and then as a small garrison for infantry and cavalry companies until its closure in 1890. After that it took on a new role as an Indian boarding school until 1959, with a brief interlude in the late 1930s as a facility for children stricken with tuberculosis.

The buildings are in exquisite condition, forming a perfect square around an open parade grounds. Gray-painted structures represent its garrison history, while the white buildings represent the boarding school past. Each structure's interior displays offer a piece of the story, from the lives of soldiers and officers to later student life in the Indian School.

Boarding school voices from Fort Totten and elsewhere call out hauntingly in the exhibits. Christine Alex, student of the Grey Nuns Indian Industrial School in nearby St. Michael, ND, lamented: "Every evening was the worst, I think, 'cause it was always lonesome. Other girls would be crying and I cried a lot of times too."[21] The Fort Totten Indian School was overcrowded with over four hundred students, and each dormitory housed up to seventy students in a single room. The school day was divided between academic lessons and industrial work in factory-like conditions, teaching seamstress and tailoring work, shoe making, farming, printing, and carpentry. Indian students were not allowed to speak their native language, only English. Alvina Alberts, a student from 1918–1926, said that discipline consisted of "pulling your ears, pulling your hair, pushing you, slapping you in the mouth if you said the wrong word."[22]

I think I understand now why the wind blew so fiercely.

---

At the Fort Totten State Historic Site museum, an oral exhibit spoke of the Seven Directions in the Dakota way of life. I thought of these in relation to our North Dakota visit. Dianne and I had been north to Pembina near the Canadian border. To the east we had followed the Red River.

Skyward had taken us up the Pembina Escarpment. Down had brought us to a historic bridge straddling the Pembina River. Far to the west, beyond our travels, the land would begin slanting upward toward the Rocky Mountains. South would soon lead us to Valley City and past the Sheyenne National Grassland.

That left the Seventh Direction, the hardest to know and understand, according to the speaker in the oral exhibit. The Seventh Direction is to the interior, to the heart. We'd dashed through the state all too quickly, I knew, but I think we'd begun this inward journey as well. In these few short days, we'd come to love, treasure, and respect this wide open space of North Dakota, one too often dismissed.

As we neared the state border it became clear as light: the Seventh Direction—the heart—is where the sky and soil, the trees and the prairie grasses, the flatlands and the hills, the bison and all fellow humans find the meeting place.

# 10

# South Dakota

## *Glacial Lakes and the Eastern Corridor: Transition and Edge*

Dianne and I had not long crossed the state line from North Dakota into northeast South Dakota when we began to notice something more. More, of course, is a relative term, and the transition was slight, maybe even a figment of our imaginations, but there seemed now to be a few more farm homes, a few more cars on the interstate, a few more row crops, and a few more lakes dotting the landscape. Even so, the differences were slight. Statewide, the population density of South Dakota is 11.3 persons per square mile compared with North Dakota's 10.7.

We were in a transition zone, though, moving from the lightly populated northeast to the more populous southeast corner of South Dakota. We were near the edge of the eastern tallgrass prairie region and the edge of row-crop farming: west of here precipitation dries up and there is more shortgrass prairie and ranching. Bird species were at a crossroads here, too, between east and west.

Edge and transition: the South Dakota landscape is where the appearance of permanence is a mirage.

Northeast South Dakota is known as the Glacial Lakes and Prairies region. No surprise on the prairies—isn't that what South Dakota is supposed to have?—although only 2 percent of the eastern half of the state's tallgrass prairie still remains. But it's the preponderance of natural ponds and lakes dotting the northeast corner's landscape that surprises the visitor. These blue eyes in the midst of corn and soybean fields and brome grass pastures keep watch on a big sky that can be benevolent or threatening, depending on the whims of the day.

On a wind-free day, I imagine these ponds reflecting the clouds, making doubles of the ducks and Canada geese foraging the waters. But wind is a near-constant here, and a kettle pond at the edge of the Waubay National Wildlife Refuge was rippled and dimpled, enveloped by watery plants like wild celery, duckweed, and duck potato, and ringed by restored prairie.

The kettle pond—the name given to small ponds formed when melting glacial chunks calved off the retreating ice and pressed down into the soft, newly exposed soil—is just the teaser. The refuge includes 4,650 acres, much of it encompassing Waubay Lake. Waubay Lake in its entirety, extending beyond the refuge, stretches across fifteen thousand acres, with numerous bays rounding off its edges and mimicking the shape of a cumulus cloud. A long causeway links the main road to the refuge headquarters located on an island in the eastern quarter of Waubay Lake.

The lake name is derived from the Lakota "Wa-bay," where the birds nest, and the combination of lakes, edge marshes, prairie, and woodlands attracts 100 nesting species and another 150 migrating species.

A trail hike around the island begins at the water's edge. The lake today is at a decades-long high level. While the nation's west and southwest suffer from drought, climate change seemed to be bringing, on average, increased rain to the Midwest. A line of denuded tree trunks extending into the lake marks an old boundary between lakes that have now fused together. Drowned trunks also line sections of lake shore, marking an old, submerged shoreline. Ironically, the higher lake levels threaten the wetland habitat. Good wetland habitat needs to dry out at its edges from time to time, giving plants a foothold to take root and provide food for nesting and migrating birds.

A boardwalk trail spans across marshes at the edge of the lake, many of them rich in reeds and rushes. The woodland trail heads into hardwood territory, dominated by burr oaks, and then marches into open prairie.

The combination of habitats and precise location makes the Waubay Refuge a transitional place. The burr oaks (reestablished since the refuge began in the 1930s after having been extensively logged) are at their westernmost fringe. Eastern forest birds are at the extent of their range, and mingle with species at their southern, northern, and eastern edges. Bird chatter in the heat of midday also suggested this must be a meeting place of sorts.

The Glacial Lakes and Prairies region of northeast South Dakota is part of a more extensive Prairie Pothole Region extending across parts of the Dakotas, western Minnesota, and eastern Montana. This region breeds nearly half of the continental US duck population.

As numerous as the lakes and ponds of the region are, though, many more were drained to increase farm acreage in the early decades of mechanized agriculture. Conservation efforts to preserve the ponds began after the 1949 publication of an article titled "Goodbye Potholes" in *Field and Stream*. Prior to that, the Soil Conservation Service had actually subsidized the draining of the pothole ponds.

More edges as we edge our way downstate. Two long, narrow lakes form part of the boundary between northeast South Dakota and western Minnesota. The first, Lake Traverse, is drained by the north-flowing Bois de Sioux River, which empties into the Red River of the North and eventually into Hudson Bay. The second, Stone Lake, is drained by the south-flowing, fledgling Minnesota River, which empties into the Mississippi and eventually into the Gulf of Mexico. Only four miles and a modest incline that hosts a segment of the Laurentian Divide separates the two lakes whose waters travel to opposite ends of the North American continent. Here the water flows northward, and there it heads to the south.

South Dakota authors write often about this landscape as a place of transition. In *Dakota: A Spiritual Geography*, Kathleen Norris writes of the Dakotas generally: "We are at the point of transition between East and West in America, geographically and psychically isolated from either coast, and unlike either the Midwest or the desert West."[1] And even within the two states, she depicts an east-west transition: "The eastern regions of both states have more in common with each other than with the area west of the Missouri, colloquially called the 'West River.'"[2] Some pundits claim the two states should have been established as East Dakota and West Dakota, rather than North and South.

Horse rancher/breeder Ann Daum, author of *The Prairie in Her Eyes*, echoes the point as she describes her family's farm where she grew up, straddling the mid-state edge between agriculture and ranching: "We straddle South Dakota's east-west split between farm and grassland. Most

Northeast South Dakota is known as the Glacial Lakes and Prairies region. Large blocks of glacial ice calved from the retreating glaciers, pressing down into the still-soft glacial outwash soil, its depressions becoming lakes and ponds that remained after the glaciers disappeared.

of the landowners [east of us] focus on farming, running a small herd of black white-faced cows to graze the land not fit for wheat or alfalfa. Just west of us, few acres are plowed for dryland wheat, and the herds of cattle and horses keep getting bigger, the range more open, all the way west to the Black Hills."[3]

If there is one thing that links east and west in the Dakotas, it is a big blue sky that stretches across unbroken horizons like an inverted ocean.

Benevolent but dangerous upon a whim, this big sky, writes Norris, "is where angels drown."[4]

Sica [*She-cha*] Hollow State Park lies just south of the boundary between the Dakotas. The Coteau des Prairies, or Prairie Hills, begin just north of here. The hills are fresh, so to speak, at this edge of the Coteau, and another transition begins as springs bubble up from below ground and emerge onto the landscape. Along the Spirit Trail, we climbed one hundred feet up a wooden staircase that wraps around the top of a gurgling seepage. On the descent I tried to pick out the precise point—the edge—where the trickling stream emerged out of the soil. There wasn't a discernible spot. Here on the slope the soil was reasonably dry, and there below it the soil was damp, and just a glance further downhill water was dribbling, joining with other rivulets, and then flowing into modest little nine-inch waterfalls before joining Roy Creek.

There are few sounds more uplifting than trickling water, except perhaps the sound of kids playing in creek water. At the trailhead, we had met a mother of a young girl and boy who had brought her kids to Roy Creek to putter about in the water, as she had done as a child. The kids' laughter followed us down the trail to the water seeps.

Ironically, though, it's not pleasant sounds that Sica Hollow is known for. In spring, trapped gasses in the thawing bog near the creek emit a low moaning as they are slowly released, leading early inhabitants to think the land was haunted. At the same time, with plentiful water, Sica Hollow was a perfect place to settle, but at a price to be paid in the form of the annual anguished voices emanating from the bog.

People seek another kind of transition when they pull into the parking lot at Abbey of the Hills Inn & Retreat Center. The center occupies the buildings and grounds of the former Blue Cloud Abbey, a Benedictine monastery that once graced northeast South Dakota.

Forty founding monks came to the Glacial Lakes and Prairies region from Saint Meinrad Archabbey in Indiana in 1950 with a mission of ministering to local Native Americans. Saint Meinrad Archabbey had been sending Benedictine priests to four different North and South Dakota missions since the late 1800s and decided the time had come to establish a new monastery closer to this particular outreach. Blue Cloud would be centrally located, lying close to the North and South Dakota border.

The monks supported themselves through the years by farming and by offering retreat space to groups and individuals. But sixty years later, only fourteen monks remained at the monastery, many of them elderly. So in 2012, the monks voted to close Blue Cloud, carrying their memories with them but leaving their deceased fellow brothers in the abbey's cemetery. Most moved on to other monasteries.

Father Thomas Hillenbrand, OSB, was a monk at Blue Cloud for most of the monastery's existence and was abbot from 1992–2007. Father Thomas joined Blue Cloud in 1959 fresh out of undergraduate college and began studying to become a priest. "One thing I liked about the abbey," he said, "was that there was not a great distinction made between the priests and lay brothers. We all worked together." Work is an essential part of the Benedictine tradition, and the Blue Cloud monks labored seventeen years to construct the abbey by hand. As a seminarian, Thomas and fellow priests-in-training studied theology from August through May but spent two afternoons a week plus summers working alongside the other monks and lay brothers, pouring concrete and finishing rooms. "I thought the manual work was just as important as intellectual study," he opined.[5]

In joining Blue Cloud Abbey, Fr. Thomas was deeply attracted to the prairie. "There's something about the plains," he noted. "It's quiet and calming. You feel a part of the soil. There's an ocean of grass rolling around you. People here are down to earth."

Benedictines have always considered themselves stewards of the land, Fr. Thomas informed me. "While the Benedictine Rule doesn't specifically mention nature, it says that monks are to 'regard the tools of the abbey as sacred.'" That means everything about the monastery is to be treated with respect: farm tools, laundry, kitchen utensils, and—of course—the land itself.

Life at the abbey was as regular as the prairie sun. The monks would rise around five or six in the morning, have Mass, then breakfast around seven. Monks studying to become priests would then go to classes while others set about their daily work. Then came lunch, rest, and a return to work or classes. Late afternoon community prayer called Vespers brought an end to the workday, followed by supper, a recreation hour (handball, swimming in the monastery's lakes, card-playing). Compline—the evening prayer—brought the community's day to an end. Monks returned to their individual rooms by 8 p.m., where most would read and study. Lights out at 10 p.m. In this way, the monks lived out the communal Benedictine way of life: prayer, work, and study.

The number of Blue Cloud monks peaked at around seventy in the early 1960s. When dwindling numbers led the monks to vote to close the abbey, Fr. Thomas likened it to "a spiritual tsunami. For a while I felt like I had been washed out to sea. It felt like losing your home and your family at a moment's notice." Within a few months of the decision, the monastery was closed, and the monks packed and dispersed around the country. Fr.

Thomas chose to join Christ the King Priory near Schuyler, Nebraska, in part to keep his feet in the prairie.

Within a year of Blue Cloud's closing, a group of local couples purchased the abbey and repurposed it as a nonprofit known as the Abbey of the Hills, an ecumenical retreat center dedicated to the rediscovery of peace.

The abbey gleamed in a hot blue sky as Dianne and I turned off Highway 12 and drove up the dusty gravel road. The former monastery's sandstone walls speak of the ages, of the Benedictine tradition that dates back to the sixth century, but its angular architecture bespeaks its 1950s construction. Dianne and I live near New Melleray Abbey outside of Dubuque, Iowa, a still-operating Trappist monastery dating back to the 1840s. New Melleray's 1870s-constructed abbey has a more traditional Gothic architecture, a look that to me had always said "monastery."

But once inside Abbey of the Hills, we found familiarity. The chapel retains its monastic seating, with rows of worship stalls facing each other from either side of the altar. In this manner, monks have chanted psalms to each other since medieval times. Invited to explore the abbey's halls and corridors, we found innumerable guest rooms. In some wings, these have been converted from monks' rooms—or cells, as they are traditionally called—but others may already have been guest rooms, as Benedictines have always been called to "greet the guest as Christ."

Later, we walked the eighty-acre grounds, letting the heat of the sun (it was ninety degrees Fahrenheit) bake our shoulders and calves while the mown grass of the prairie trails warmed our sandaled feet. The trails around the upper and lower lakes reminded me of Glendalough, my favorite Irish sixth-century monastic ruins (the name Glendalough meaning "the valley of the two lakes"). Another path took us along an outdoor Stations of the Cross. We sat for a while on a shaded bench before a statue of Mary.

Only the plastic kayaks lying along the upper lake's beach betrayed a postmonastic presence.

Deacon Paul Treinen is director of Abbey of the Hills, but the center is at heart a collaborative effort. "Six of us couples were having dinner several years ago," he recalled, "and the talk turned to Blue Cloud Abbey shutting down. We wondered aloud what would happen to the Abbey as it had been for sale for about a year, when out of nowhere someone said, 'Maybe *we* should buy the Abbey.'"[6] The idea took hold, though none of them knew yet what they would do with the abbey, and there was already an offer on the table. The couples almost backed away, but at Mass on the morning of the day they needed to make a decision they kept looking at each other and nodding their heads. "The Holy Spirit was telling us to put an offer in—to trust him and put an offer in on the Abbey," Paul said. That morning, Paul called the abbot, who initially dismissed him, saying, "We're only taking serious offers at this time."

Paul's reply: "I think we *are* serious."

Eventually, the six couples' offer was accepted. They still didn't know exactly for what purpose they would use the abbey, only that they wanted to maintain it as hallowed ground, as a place to foster the rediscovery of peace. One of the more difficult things for the monks to resolve as they pondered the abbey's closing, Paul related, was how their commitment to the land would continue in their absence. The Abbey of the Hills' vision seemed in step with the monks' wishes. The land itself would become part of their mission of peace.

Paul is intimately connected to the landscape. On the morning we talked, he had arisen at four thirty to bicycle twenty-eight miles through the rolling hills and arrive at the abbey by seven thirty. "This area I live in has most recently been affected by glacial activity," Paul said. "And before that, it was part of Glacial Lake Agassiz. It reminds me of the bigness of God, of the timelessness of God," through the slow changes of the landscape and the evolution of one's own personal growth.

The group slowly came to realize that the abbey should be utilized as an ecumenical retreat center. They have hosted Lutheran minister retreats, Methodist retreats, Seventh-day Adventist gatherings, Catholic priest and deacon retreats, veterans' retreats, twelve-step program retreats, family gatherings, farm management gatherings, and individuals who simply come to the abbey "to rediscover peace."

The abbey's rural location—two hours from Fargo and three from the Twin Cities—aids in its mission. "When you get away from light pollution, when you experience the prairie silence, that can be a compelling place for a soul to listen to God," Paul said.

Kathleen Norris equates the Dakota landscape itself with the monastic experience. "Nature, in Dakota," she writes, "can indeed be an experience of the holy," but it "requires you to wrestle with it before it bestows a blessing." The Plains, she continues, "seem bountiful in their emptiness, offering solitude and room to grow."[7]

Brother Benet Tvedten, OSB, a deceased monk and author from the former Blue Cloud Abbey, pondered the transformative nature of living in these hilltop plains:

I can understand why St. Benedict left the city. Sometimes when I look at the lights in the valley, I think of things I'd like to do in that imaginary city. Most of the time, though, I am satisfied to be where I am. And at dawn when we are on our way to morning prayer, the lights in the valley are going off and the sun is rising, and then I see the reality—hay bales and fields of corn and alfalfa. It is much better for a monk to live in the country. . . . There is no illusion here. This is where our bodies wait for their resurrection.[8]

Dianne and I are soon back on the road again, scouting out the countryside. The gravestones in a rural cemetery suggest immigrant German farmers: Peter Linduer, 1826–1903; Max Graen, 1867–1905; Joseph Graen, Geb 1861–Gest 1885. These were farmers who had inherited a land ill-gotten from indigenous peoples and who had by their hard-earned sweat turned the prairie into cropland and founded many a rural town. The road took us to Kranzburg and Strandburg and Estelline and Rauville.

The countryside is an open book of agricultural change. The South Dakota State University (Brookings, SD) Museum of Agriculture logs a succession of change and transition on the rural landscape, starting with horse-drawn plows and horses on treadmills powering threshers. Steam brought powered tractors and railroads, and with the railroads came grain elevators that still dot the landscape for storage and shipping of wheat and corn. In the 1890s, populism brought farmer co-ops collectively owning grain elevators and working collaboratively to maximize sales; 1930s co-ops brought electrification to the rural countryside. Diesel engines brought more reliable, larger tractors, and the digital age has brought GPS-precision planting, fertilizing, and pest control.

Farm economics has made farms fewer and larger. In 1940, South Dakota boasted 72,000 farms at an average size of 540 acres. By 2019, there were only 31,000 farms at 1,460 average acreage.[9]

Wind turbine farms grow like corn on steroids today in this wind-swept state. South of Watertown, we counted seventy-three turbines in a horizon-to-horizon visual sweep.

Waves of cultural conservatism have changed the landscape as well. Today there are vastly fewer co-ops as farmers slog it out individually. A rural road sign grouped several grievances together in a single proclamation: "Eat Meat, Wear Fur, Keep Your Guns: The American Way."

Many a small town has suffered through recent decades, a fate not unusual in the rural Midwest or anywhere for that matter, plagued by closed storefronts and homes empty or in need of repair. But Watertown, SD, is one of those medium-sized towns that has escaped economic blight. In fact, it has achieved a remarkable record of growth. Its population has increased almost continually through the decades, having almost doubled in fifty years from thirteen thousand residents in 1970 to twenty-three thousand in 2020. Scott Wahl, Memorial Park campground director on Lake Kampeska in the northwest corner of Watertown, attributes Watertown's success in part to a group of business and community leaders called Watertown 2020 who had set out in prior years to envision what they might want their town to look like. The plan they enacted included educational links between the local, nationally ranked technical college and local businesses, upgrades of parks, a new senior citizen center, and improvements to the airport. The town's economic engine is driven to a large degree by Terex Corporation, which builds trucks and makes trucking supplies. The town is currently building a new ice-skating rink. The Terry Redlin Art Center is a tourist draw for fans of his realistic nature and rural life paintings. "We enjoy being a small town," said Wahl, "but we want to have a little growth."[10]

The present is built on a solid past in Watertown. Dianne and I took a self-guided walking tour through town to locate old treasures such as the 1880s Italianate villa Mellette House, built by the first governor of South Dakota; the Queen Anne-style 1887 Mathiesen House built by Norwegian immigrants who ran a successful stock farm; and the Prairie School–style 1928 Melham House built by the owners of the Watertown Sash & Door Company.[11] Fifty-six houses grace the walking tour, harkening back to the town's early success.

Dianne and I were particularly drawn to the fourteen-mile bike trail that loops Lake Kampeska and its offshoot trails that lead into town. We camped two nights on the western shore at Memorial Park, lulled to sleep by the waves that pick up bulk after two miles of rolling across the lake.

Wahl, who has been campground director for three years after retiring from food service in the Watertown school district, notes the mesmerizing effect of the lake. "I see people sitting quietly by themselves at the lake's edge every day. I'd like to ask them what they're thinking, but I respect their privacy," he said.

It took some changes to restore Lake Kampeska's water quality after years of abuse. In the 1880s, Watertown began drawing its drinking water from and discharging its wastes into the lake. Over the decades, over seven hundred homes were built along the lake's perimeter with individual septic tanks and outhouses. Agricultural and urban sedimentation reduced the lake's average depth from twenty-five to fourteen feet. But in recent years, centralized sewage treatment has lessened the pollution of the lake, the city has supplemented its water supply with deep wells, and limits have been placed on new lakeside housing. These and other changes have helped Lake Kampeska to reasonably recover.[12]

Find of the week: the Dakota Nature Park just south of Brookings, a 135-acre park with bicycle and walking trails, fishing ponds, and prairies. The trails and prairies were developed over a landfill that was capped and closed in 1994. The ponds were gravel excavation pits that filled from the water table below.

From prairie to landfill to prairie. Another kind of transition.

The eastern corridor of South Dakota changed as we drove south. The tranquility of glacial lakes and ponds gives way to local streams and rivers that have cut viciously into the landscape, exposing a Sioux quartzite bedrock. At Devil's Gulch, a small but active stream still eats away at the pinkish rock, forming a narrow, deepening canyon. Devil's Gulch gained fame in 1876 when the outlaw Jesse James and his horse took a twenty-foot leap across the gorge, eluding capture after a wild pursuit following a bank robbery 170 miles away in Northfield, MN.

From Devil's Gulch, Split Rock Creek continues about three miles downstream to Palisades State Park, where the quartzite pokes through the landscape again. The creek—larger now—exposes and bisects fifty-foot cliffs and towers of the pink, angular rock. Crevassed vertically, etched horizontally, occasionally leaning, and topped by boxy blocks of stone, the palisades look like small, precariously perched skyscrapers. Dianne and I scrambled over the rocks like salamanders, clinging from stone to stone while climbing onto ledges and descending again to creek side.

On the southern end of the park, a trail descends through a breach in the rock and empties onto a grassy shoreline path where the quartzite lay ten or twenty feet back from the creek's edge. Here the quartzite begins to recede from the visible landscape. Back there is fifty feet of cliff and tower, then twenty, ten, and here—right here—the quartzite disappears into the ground.

The rock is 1.8 billion years old, laid down as a sandy sea bottom. Pressurized by weight and movement of the continental plate, it slowly metamorphosed into Sioux quartzite. In places we found ripple marks on flat exposures offering a glimpse into its sea-bottom past.

Split Rock Creek has carved through the underlying quartzite bedrock at Palisades State Park in southeast South Dakota.

Periodically, the South Dakota sea bottom would be inundated with river mud, resulting in small seams of shale, mudstone, and pipestone within the prevailing quartzite. Native Americans mined the soft pipestone here and where the formation continues into southwest Minnesota and shaped it into ceremonial pipes.

The Big Sioux River takes its own turn slicing away into the quartzite twenty-six miles farther southwest in Sioux Falls. Once buried under sediment, the quartzite here was exposed at Falls Parks when the Big Sioux long ago carried a rush of glacial meltwater. Today the river appears to have adopted a divide-and-conquer strategy, splitting right and left and nosing out the weak spots in the rock, through which it spills in numerous small and larger waterfalls through the course of a hundred-foot drop. Along with a throng of other visitors, we milled about the rocks, each of us edging up to our own personal favorite falls.

There is a "this-ness" to creation, a point at which the immensity of nature comes down to one particular place. The Sioux quartzite sits at surface level in southwest Minnesota through southeast South Dakota and lies buried through the center of the state and on into Nebraska. In places it is up to four thousand feet thick. But subsurface transitions to surface, and all that vastness whittles eventually to one small falls where part of the Big Sioux River squeezes through a gap in the quartzite at the base of my feet.

We walk on ancestral lands.

Nothing about that transition from indigenous to largely Euro-American occupation was pretty or gallant, yet here we are. And here I am. The least we can do is to try to remember.

Of course we need to remember the poison treaties and the forced relocation of Native American tribes. In eastern South Dakota, the 1858 Yankton Treaty ceded 11 million acres to the US government due to relentless White immigration and relocated the Sioux to the Yankton Reservation along the Nebraska border. A Dakota treaty signer, Padaniapapi (Struck-by-The-Ree), despaired, "The white men are coming in like maggots. It is useless to resist them. They are many more than we are. We could not hope to stop them."[13]

Equally important is to remember and to celebrate the indigenous civilizations that graced the continent prior to Euro-American contact.

Dianne and I pulled into Good Earth State Park on the morning of our last full day in South Dakota. Located about eleven miles southeast of Sioux Falls on the hilly banks of the Big Sioux River that marks the boundary with northwest Iowa, Good Earth was the site of a flourishing intertribal city that lined both banks of the river from AD 1500 to 1700.

This was our own personal learning curve. In our home corner of northeast Iowa, we have become well versed in the Woodland tradition that produced the Effigy Mounds and other mounds that line the Upper Mississippi River dating as far back as 500 BC. We've made ourselves familiar with the story of Black Hawk and his ill-fated attempt to reoccupy the Sauk tribal home on the east side of the Mississippi, and the so-called Black Hawk War of 1832 that left nearly fourteen hundred Sauk dead from starvation, skirmishes, and massacre. We've learned about the Meskwaki removal and return to Tama, Iowa. We've learned the southwest Wisconsin story of Ho-Chunk removal and return.

Farther west, for Dianne and myself, the story is more of a jigsaw puzzle, one that we'd only begun to piece together. Good Earth provided us with a solid corner from which to begin.

The newness and freshness of the site was apparent as we arrived. South Dakota's newest state park—dedicated to education, celebration, and remembrance rather than to RV camping—the six-hundred-acre Good Earth State Park opened in 2013. Its visitors' center museum and hiking trail kiosks tell the story of the city that flourished here. Other people are learning, too.

Something in the landscape itself speaks of transition. In their book, *Blood Run: The "Silent City,"* Dale Henning and Gerald Schnepf notice immediately that "the wide valley that carries the Big Sioux River, separating Iowa from South Dakota, gives you the impression that this is where the West begins." But beneath the soil another story is told, that of the ancestral Omaha, Ponca, Ioway, Otto, and other occasional tribes who lived here together in peace and prosperity for about two hundred years, dispersing around 1700, prior to the arrival of Euro-Americans.[14]

Rather than a single, uninterrupted city, however, Good Earth may have looked more like a series of closely neighboring villages. Altogether its population likely exceeded six thousand residents living in circular and oval-shaped wooden lodges. Archaeological studies have discovered seventy-six lodge outlines, their existence evidenced by the stones outlining their edges and which anchored the hides and mats that covered the lodges. Circular lodges ranged from twelve to forty-eight feet in diameter, and oval lodges ranged from sixty to one hundred twenty-three feet long. Villagers constructed burial and ceremonial mounds near the lodges—at least eighty of which have been identified out of an estimated 275. A great serpent mound, one-eighth of a mile long, is believed to have existed on the site but was largely obliterated by an 1800s railroad line and local agricultural plowing.

Excavated refuse pits show a diet of bison, deer, and dogs, as well as corn and squash. Arrow points, stone mauls, shaved-antler hide scrapers, and two-handled clay pots tempered with ground-up shells were among the cooking and serving utensils discovered on-site.

Good Earth was a prosperous trade center as well. Its location on the Big Sioux River, about seventy-five river miles north of its confluence with the Missouri, gave it trade access with Native Americans across the North American continent. Trade artifacts found at Good Earth include obsidian from Wyoming and Idaho, chert from southeast Nebraska, and copper artworks from Upper Peninsula, Michigan. Among other goods, local artisans made pipes, intricately etched decorative tablets, and other objects from the pipestone mined nearby that they in turn traded.

Although the Good Earth village existed prior to the region's first contact with Euro-Americans, signs of change were in the air even as Good Earth thrived, not to mention change among the traded goods. Arriving tribal merchants brought with them Jesuit rings and Dutch marine shell beads called "runtees" that signaled eastern tribes' contact with explorers, missionaries, and colonists.

By the early 1700s, European colonial expansion in today's eastern US and Canada was creating a ripple effect. Eastern tribes were being forced westward. Pressure from the plains Sioux as they, too, sought new lands likely caused the Omaha, Ponca, Ioway, and the Otoe to abandon Good Earth. The Ioway moved south along the Missouri River in Iowa and Missouri, the Otoe to the Platte River in Nebraska, and the Omaha and Ponca to southeast and northern Nebraska, respectively.

A few early archaeologists took notice and made records of the abandoned site in the 1880s, but for the most part, it was swallowed up by pioneer agriculture and the Burlington, Cedar Rapids, and Northern Railway. By 1970, enough modern interest had coalesced around the site that it was declared the Blood Run National Historic Landmark (Blood Run being the name of a creek on the Iowa side of the village). In 2013, Good Earth became a South Dakota state park.

Good Earth, of course, is a modern name given to the village, whose actual name is unknown. But the name was created in consultation with the ancestral tribes.

Dianne and I ate lunch at a pavilion overlooking the prairie and then walked the grounds. The grass paths were warm against our sandaled feet on this hot summer day. Restored prairie blazed with wildflowers on the uplands, and woods outlined the steep ravines and valleys. At an observation deck we looked out over the Big Sioux and its wide valley, across to the farms and silos on the horizon, and thought about the changes that had come across this land.

We found a hotel in Sioux Falls for our last night in South Dakota to freshen up after several nights of tent-camping and to start well rested for home in the morning. By ten o'clock the next morning, we had crossed over into Iowa at Sioux City and took a brief detour to a bluff overlooking the confluence where the Big Sioux River joins the Missouri.

Here we were at the place where east meets west, the heart of the Midwest. You might call it a divide, an edge, or you might look out over a space that seems to transition only gradually across the landscape. You might think of a geology and history that sometimes changes abruptly and catastrophically, or of deep time that is the long, slow slide of endless sunrises and sunsets.

We were out of time, for this trip at least, and so we drove down off the bluff and rejoined Highway 20, the road that would transition us east across Iowa, through the hilly western border, across the flat central plain, to the unglaciated hills of the Driftless. Highway 20 would bring us within two blocks of home.

# 11

# Minnesota

## *Lake Itasca, Headwaters of the Mississippi River: All One Thing*

A thin January sun spills through the pines as Itasca State Park supervisor Robert Chance shuttles my friend Andy and me through the northern Minnesota park, famous as the headwaters of the Mississippi River.[1] The roadsides are heaped with freshly plowed snow. The air is a crisp but hardly intense ten degrees Fahrenheit above zero.

At a boat landing not far from the headwaters, Robert swung the SUV around and mused, "These places that we find desirable have been desired by others before us." He was recalling the occasion—years earlier in his Minnesota Department of Natural Resources (DNR) career—when his vision of adding public access to a remote lake ran headlong into the archaeological remains of a Native American village at the site. But he could just as easily have been referring to Itasca itself, where the layering of landscape, the Mississippi River, and the human story are thickly intertwined.

A drive through the North Country where Lake Itasca resides quickly reveals that we're not in someone's old, tired impression of the Midwest. Here in northern Minnesota the landscape is pocked with glacial lakes hidden behind deep swaths of Northwoods pines or poking out from wide stretches where farms have replaced the prairie. The farms are less dense here, though, than in much of the Midwest, sharing the landscape with reeds and wetlands and long stretches without homes.

Andy and I had come to Itasca from Dubuque, Iowa, where the Mississippi passes eight hundred river miles downstream from the headwaters and sixteen hundred miles before it spills into the Gulf of Mexico. Here, too, the human story is tied tightly to the great river.

My home in Dubuque lies in the Driftless Area, another unique region of the Upper Midwest. The land is called "Driftless" because there is no glacial drift here—the glaciers repeatedly bypassed and even encircled the landscape, leaving it rugged amid the glacially sculpted Midwest. With its rugged, steep valleys, rock towers, and river bluffs, the Driftless has always struck me as a region of mystery. But I'd always thought that this great river must link as well to another place of equal, if different, mystery at its North Country source. Both the Driftless and the Headwaters, it seemed to me, lay in contrast to the straight, corn-rowed plains of the greater Midwest.

We'd both been to Lake Itasca before (Andy is Robert's brother-in-law), and in more inviting summer weather, but the timing worked out for a couple of teachers on semester break, so we loaded our snowshoes and overnight bags into my car and set off on a January road trip to the Mississippi headwaters.

Itasca is a thousand-acre, northern Minnesota lake shaped like a wishbone, an inverted Y, with two lobes pronging off to the south. Or it's like inverted antlers, which may have led to its Ojibwe name, *Omushkos*, or Elk Lake in English. It was called Lac Le Biche by the French before it was renamed Itasca.

A fledgling Mississippi River pours from the north end of the lake through a thirty-foot, boulder-strewn outlet. The river here is knee deep, clear, and achingly cold in contrast to the giant that lumbers through the Dritftless. Canoeists who ply their way from the Headwaters to the Gulf of Mexico will, near Itasca, bump their crafts' noses on either side of the shore on any of the numerous bends in the river and may find themselves tugging their boats over the shallows. The river will grow by leaps and bounds from here. Lake Itasca lies twenty-four hundred miles[2] above the Gulf of Mexico. By Louisiana, the Mississippi often exceeds a mile in width and runs almost two hundred feet deep near the French Quarter of New Orleans. At the delta, it will pour 420 billion gallons of water into the Gulf of Mexico each day.[3]

Located just two hundred miles south of the Canadian border as it exits Lake Itasca, the tiny Mississippi meanders north and east for sixty river miles till it is turned aside by a three-way continental divide on the Giants

Range, a highlands plain. First the Laurentian Divide diverts waters on its north face to the Hudson Bay, then the St. Lawrence Divide diverts water east to the nearby Great Lakes and the St. Lawrence Seaway. Almost in recompense for rerouting the still-young river southward, the divides direct all waters on their southern and western flanks to the Mississippi, and on to a long journey through the Driftless and down to the Gulf of Mexico. In all, the Mississippi River, with help from the Missouri and Ohio Rivers, will drain 40 percent of the continental United States.[4]

The mystique of America's central river draws half a million tourists—domestic and international—to the headwaters each year. On a typical summer's day, young and old will totter across the slippery glacial boulders where the Mississippi pours forth. Dozens at a time wade across the pebbly bottomed bed of the infant river.

Robert's wife Kathy told us, later, that she remembers their young daughter playing in the newly birthed river with a girl her same age from another country who spoke no English. "The river was their common language," Kathy said.

Although the glaciers never reached the Driftless Area, their meltwaters carved and deepened the Mississippi Valley against a backdrop of cliff-like river bluffs. By contrast, those same mile-high glaciers deposited a maze of hills and ridges onto the Itasca landscape. The repeated advances and retreats of the mile-high ice left behind moraines, kames, eskers, and drumlins formed when the glaciers dropped their payloads of soil, sand, pebbly rubble, and well-polished boulders in their final melting retreat twelve thousand years ago.

The fingerprints of glaciers are everywhere in the region. A hundred marshy glacial lakes dot the thirty-three-thousand-acre Itasca State Park, and many of them are kettle ponds formed from huge ice blocks that broke off from the retreating glaciers then sunk by their great weight into the still-soft glacial outwash. South of Lake Itasca, near Park Rapids, the land flattens out again, having first been bulldozed flat beneath the advancing glacier and later overlain with outwash rubble from its meltwaters.

The Lake Itasca basin itself is a glacial tunnel valley, formed by a rushing river at the bottom of the ice that scoured out the land beneath the glacier. Today's lake covers 1,065 acres at an average depth of seventeen feet, a maximum of forty.

After the Mississippi River emerges from its headwaters at Lake Itasca, Minnesota, it briefly flows northeast before the continental divide turns it southward along its twenty-three-hundred-mile route to the Gulf of Mexico. The Mississippi watershed drains 40 percent of the continental US, plus a small area of Canada.

I am a lover of textured, layered landscapes and the stories of the peoples who have settled there. From the Mississippi bluffs near my home I ponder sunrise mists loitering just above the river. Elsewhere in the Driftless, rock towers rear up unexpectedly where a country road descends and twists into a stream valley. Springs erupt mysteriously from the bedrock and disappear again into the karst landscape.

The human story in the Driftless is deeply connected to the landscape and the Mississippi River. Native Americans first called the Driftless Area home twelve thousand years ago as they hunted just south of the glaciers. Later, Woodland period peoples built burial mounds, many in the shape of great bears and birds, above the Mississippi. The Meskwaki, Sauk, and Ho-Chunk witnessed the arrival of French-Canadian trappers and miners such as my own town's Julien Dubuque, who came to mine lead in 1788. The 1832 massacre of the Sauk on the Mississippi, ending the so-called Black Hawk War, brought Euro-Americans, who established cities, towns, and farms along the river.

The human story at Itasca followed similar patterns, lagging somewhat in time as the ice retreated. Ancient peoples followed the melting glaciers northward, hunting megafauna just beyond the retreating ice. An eighty-five-hundred-year-old bison kill site in the boggy landscape near the headwaters was unearthed in 1937. At the site, archaeologists found knives, scrapers, and arrow points used by the people, as well as bison, deer, bird, and fish bones. Later peoples built burial mounds not far from the headwaters.[5]

The Ojibwe populated the region beginning in the 1500s, having migrated westward from the Great Lakes and St. Lawrence River. Supported by the French, they battled and displaced the Dakota in the Itasca region by the early 1700s.[6]

French Canadian fur trappers began arriving in the 1600s, and Euro-American intrigue for locating the headwaters of the Mississippi River followed within a century. The Paris Treaty of 1783 ending the War of Independence established the Mississippi River as the western boundary of the United States, igniting a rush to locate the headwaters and firmly establish the border.[7] But the ganglia of northern Minnesota lakes and connecting rivers, as well as the lack of an established definition of a "headwaters," made exploration of, and agreement about, the true headwaters problematic. It would prove difficult to settle on a single source.

The British-Canadian fur trader and surveyor David Thompson was among the earliest headwaters explorers, claiming Turtle Lake as the river's source in 1798.[8] But while the lake's waters flow into the Mississippi River, cartographers today call its inlet and outlet the Turtle River. The Mississippi lies south of Turtle Lake.

Two years after the 1803 Louisiana Purchase moved the national border westward, the US government commissioned General Zebulon Pike to map the Upper Mississippi River, study its plant and mineral resources, select locations for forts, and establish contacts with Native Americans. But Pike and his twenty soldiers set off too late from St. Louis in summer 1805,[9] and by the time they reached Leech Lake, about 150 miles downstream from Itasca, deep winter had set in, and his men were ill fit for further travel. Pike knew of Thompson's earlier findings but declared the waters above Leech Lake too insubstantial to be called the Mississippi and dubbed Leech Lake the headwaters instead. But Pike also took a small group of his men northward on a side-expedition and simultaneously declared Red Cedar Lake, thirty miles upstream, a "minor upper source."[10] The "source" was moving upstream and was perhaps not singular after all.

Next up was Lewis Cass, who, in 1820, retraced Pike's journey as far as Upper Red Cedar Lake (today named Cass Lake) before turning around. Unlike Pike, however, he knew he hadn't reached the headwaters, having heard that the Mississippi could be traced to Lac Le Biche.[11]

In 1832—the same year that the Sauk were massacred at the Mississippi River a few hours north of my home in the Driftless—Henry Schoolcraft, a crewman from the Cass expedition, was commissioned by the US government to visit the northern Minnesota region to settle disputes among tribes and inoculate them against smallpox. He set out toward Leech Lake with a crew of thirty-five to conduct his business. Schoolcraft made it his own separate and unannounced goal to reach the headwaters once and for all.[12]

Having completed much of his commissioned work, Schoolcraft left most of his men behind at Cass Lake in early July and set off with a smaller crew in search of the headwaters. Unlike previous explorers, Schoolcraft relied heavily on Ojibwe guides, most prominent among them Ozawindib, an Ojibwe chief. Each boat in the five-canoe flotilla also included Ojibwe guides and paddlers.[13]

As they neared Lac Le Biche,[14] Schoolcraft lamented the marshy ground wherever they had to portage: "A man who is called on for the first time, to debark, in such a place, will look about him to discover some dry spot to put his feet upon. No such spot however existed here. We stepped into rather warm pond water, with a miry bottom."[15]

Closer to Lac Le Biche, though, lay firmer ground, and Schoolcraft's mood lifted. He noted that others had once occupied this land: "The carbonaceous remains of former fires, the bones of birds and scattered camp poles, proved it to be a spot which had previous been occupied by the Indians."[16]

Portaging further, they came at last upon Lac Le Biche on July 13. The lake, Schoolcraft writes, was "in every respect, a beautiful sheet of water seven or eight miles in extent. . . . The waters are transparent and bright, and reflect a foliage produced by the elm, lynn, maple, and cherry, together with other species more abundant in northern latitudes."[17]

Finally he describes the outlet, the beginning of the Mississippi River, "perhaps ten to twelve feet broad, with an apparent depth of twelve to eighteen inches. The discharge of water appears to be copious, compared to its inlet," the lake's volume and discharge having been accentuated by numerous springs beneath its surface.[18]

On borrowed time away from his sanctioned and commissioned tasks, Schoolcraft's party quickly planted a flag on the lake's only island and held a brief service whereupon Schoolcraft renamed Lac Le Biche by combining parts of the Latin words for "truth" and "head," *verITAS CAput,* or ITASCA, the "true headwaters."[19]

The name Elk Lake was later bestowed on another lake directly upstream from Itasca, thus depriving it, too, of its Ojibwe name.

It was all one thing, this loss of land and life, and even of names.

In 1836, French geographer Joseph Nicolett poked a few holes in Schoolcraft's headwaters exploration—inlet holes, to be exact. After charting several inlet streams and upper lakes that fed into Itasca, Nicolett graciously claimed, "I come only after these gentlemen [Schoolcraft and his crew], but I may be permitted to claim some merit for having completed what was wanting for a full geographical account."[20] But he had opened the door as to whether the Mississippi could be traced to a single source.

Fifty years later, in 1881, Civil War leader and travel writer Captain Willard Glazier laid out his rationale why Itasca *wasn't* the headwaters. Glazier pointed to a small creek flowing into Itasca from an upstream lake he claimed to have found. In the tradition of headwaters explorers before him, Glazier published a narrative of his discovery and promptly named the new lake after himself. But suspicion soon arose when the public noticed Glazier's plagiarisms from Schoolcraft's text.[21] The newly discovered Glazier Lake turned out to be the already-discovered Elk Lake.[22]

Glazier eventually admitted his fraudulent writings and discoveries. But the question still remained: if Elk Lake lay upstream from Lake Itasca, which one *was* the headwaters of the Mississippi? Was there even a single source?

In 1888–1889, the Minnesota Historical Society commissioned surveyor Jacob Brower to settle once and for all the debate over the headwaters. Brower split the difference, so to speak. He acknowledged that upper lakes and streams contributed flowage to Lake Itasca and in that sense might be considered *as a region* to constitute the headwaters. But these streams and flowages were small (he called them the "infant Mississippi"), and even occasionally dried up in the heat of summer. Only the stream leaving Lake Itasca was constant, significant, and large enough to be called a *river*, and thus he declared Lake Itasca the headwaters of the Mississippi.[23]

In the end, it was western linear thinking *par excellence*—or *par obsession*—that insisted there be an identifiable, single headwaters to the Mississippi.

The Ojibwe, despite participating as guides, were overall bemused by this obsession. According to Leech Lake Reservation historian Larry Aitken, to the Ojibwe "it was amusing, that aggressive effort. . . . Because it was not important where it started—the whole river was of central importance."[24]

It was all one thing.

---

Robert, Andy, and I abandoned the relative comfort of the SUV to hike the quarter-mile trail down to the headwaters. Soon the Mississippi lay in front of us: a narrow and shallow stream tumbling energetically over glacially rounded boulders, emptying from the lake in a swath about thirty feet wide. The lake itself was iced over, but the stream bubbled with such vivacity that it flowed unobstructed for some distance.

We watched the tiny Mississippi emerge for a while. I considered edging down to the shore to stick my hand into the bubbling waters to say that I had *touched* the river that would flow past my home, but having moments earlier almost slid into the water across the ice-encrusted snow, I decided otherwise.

Robert had other things to show us, so after a short while we retraced our steps. About thirty feet downstream, Robert pivoted and showed us erosional damage occurring as the exuberant stream collided with the shoreline, the damage accentuated by millions of tourist feet over time. Robert wanted to rearrange the glacial boulders at the outlet to channel the flow more to the center. But the idea had received some flak from historical societies, he told us, which is ironic because in the 1930s, the Civilian Conservation Corps (CCC) moved the river's outlet some distance from where

Schoolcraft had encountered it to create better tourist access away from the marshes.[25] Robert pointed out a dearth of trees in the distant marsh where the Mississippi had formerly emerged. To say that *this one place* has always been the headwaters flies in the face of history.

---

The logging industry followed exploration of the headwaters almost immediately. In 1837, the US government wrested possession of northern Minnesota and northern Wisconsin from the Ojibwe, and 1838 brought the first entrepreneurs with eyes set on the expansive northern forests. The first northern Minnesota sawmill commenced ripping trunks into floatable logs the following year.[26] These logs would be bound together and floated down the Mississippi to sawmills in Driftless Area river towns like my own.

Massive old-growth pines fueled the drive, with 3.5 million acres of northern Minnesota dominated by white pines up to four hundred years old. A single such tree, according to Minnesota forester Chuck Wingard, could "build a small barn, a farmhouse, a church."[27] By 1900, over thirty thousand lumberjacks worked the northern Minnesota forests, harvesting $1 billion worth of trees annually. Although awareness was already growing that the great forests were dwindling, the industry continued felling pines, seeing deforestation as merely the first step to opening the land up to agriculture.[28]

Having established Lake Itasca as the Mississippi River headwaters, Jacob Brower and a small group of conservationists lobbied to preserve the last stands of native pines in the vicinity of the lake. In 1891, by a one-vote majority of the state legislature, Itasca became Minnesota's first state park, and the second oldest state park in the nation. But the law only provided the state with a patchwork ownership of lands surrounding the lake. It provided even fewer resources for regulating the logging industry.[29] Some areas of the intended park boundaries were still privately owned and open to logging, and the lake and river remained available to loggers to float their product. Loggers and the early park commissioners quarreled over how much the loggers could dam the lake and flood the river to float their logs downstream.

One of the most important figures in the battle with loggers was Mary Gibbs, named Interim Commissioner in 1903 at age twenty-four when her father, the park's fourth commissioner, died unexpectedly. Objecting that logging dams threatened the survival of shoreline trees by raising the water level of the lake, Gibbs famously confronted loggers who threatened to shoot if she so much as touched their headwaters dam. Gibbs retorted, "I will, too, put my hand on [the dam levers], and you will not shoot it off."[30] The loggers sat by as Gibbs raised the dam gate and lowered the water level on the lake. Gibbs's persistence established that Lake Itasca would not be destroyed for the sake of logging.

By the 1920s, all logging had ceased at Itasca State Park.

Despite—or because of—her success, Mary Gibbs's tenure as Interim Park commissioner lasted only four months before she was replaced.

A sign back at the Itasca State interpretive center reads "Come for the river, return for the pines." It is all one thing. But Robert reminded us that the pines, not the river, were the reason for the park's inception. Indeed, the park today harbors 20 percent of Minnesota's remaining old-growth forest.

We were back on the park roads again in Robert's SUV, and he commented on the current state of the forest. The few remaining old-growth red pines date roughly back to the 1700s. Along the road we saw numerous downed trees and stumps whose trunks had snapped about three feet off the ground during windstorms in 2012 and 2016. Robert explained that the blown-over trees had been weakened or already killed by age, disease, pests, and past windstorms. Foresters were reseeding with white pine where possible.

The forest also harbors deciduous trees typical of the Northwoods, like ash, aspen, and birch, as well as hardwood oak and maple. Some of these, too, are battling infestations, like the emerald ash borer. Climate change exacerbates the problems. Robert, who grew up in northeast Minnesota's Iron Range, remembers long winter stretches when night-time temperatures plummeted to minus forty degrees Fahrenheit. Such long, intense cold spells were integral in knocking back pest populations.

The Mississippi River had already become a highway of floating logs before Itasca began harvesting its forests. By 1836, the first of several steam sawmills began operating in Dubuque to cut the logs that had been floated downstream in massive lumber rafts.

And while Mary Gibbs put a stop to a logging dam that endangered the Itasca shoreline, the downstream Mississippi River would later be harnessed with twenty-seven locks and dams from the Twin Cities to southern Illinois in a 1930s Works Progress Administration (WPA) project to ensure a nine-foot-deep channel for barges. Lock and Dam #11 stretches across the Mississippi River at Dubuque.

The environmental impact of the lock and dam system is complicated. A typical fifteen-barge tow can move as much grain as a thousand semitrucks. But large amounts of silt are trapped behind the dams, slowly filling up backwater habitats.

Then again, in winter the agitation below the dam gates provides open water that attracts bald eagles scoping for fish in the otherwise frozen river.

"This is wolf country," Robert said, back on the road at Itasca. A wolf pack is known to live on the southern end of the park. Its alpha male occasionally strays south of the park, but not too far, because another pack's boundary lies nearby. Similarly, a wolf pack to the north avoids the park. "It's as if they've adopted the park boundaries as their own," Robert marveled.

The vast Itasca forest, bog, and meadow habitats harbor a northern Midwest wildlife oasis. A 2003 publication of the Minnesota Department of Natural Resources listed *canis lupis*—the Gray Wolf—as "Common,"[31] not so many years after a 1959 report had indicated the once-prolific species was "no longer present in the park."[32] In addition to the usual northlands roster of mammals (shrews, rabbits, rodents, beavers, raccoons, foxes, coyotes, squirrels, and the like), Itasca is home to black bears and minks ("Common"), and to bobcats and mountain lions ("Rare"). The Woodland Jumping Mouse is "Rare," but would be really cool to see.[33]

The quality of lake habitat is under duress, however. Upstream from Lake Itasca, Elk Lake may finally be getting its due attention—from biologists, at least—after having been denied status as the headwaters. Elk Lake is one of twenty-four "sentinel lakes" scattered throughout Minnesota being studied to determine "how major drivers of change such as development, agriculture, climate change, and invasive species, can affect lake habitats. Elk Lake was selected to represent a deep, mesotrophic lake in the Northern Lakes and Forests (NLF) ecoregion. With the exception of a group campsite and a paved nature trail, there is no development on Elk Lake."[34] In addition, "approximately two-thirds of Elk Lake's watershed was never logged."[35]

That said, the health of Elk Lake is still at risk. Despite not being directly touched by human habitation, it hasn't escaped the latent effects of humans on the environment at large. Elk Lake saw an increase in lake temperature of one to two degrees Celsius from 1985 to 2010, with resultant declines in native fish species.[36]

Nearly lost amid the fifty-four-page biological study is a quick line stating "[Elk Lake] represents the headwaters of the Mississippi River."[37]

---

The river that exits the North Country develops a whole new set of environmental issues as it lumbers through the rest of the Midwest and beyond. The Mississippi is much maligned for carrying erosional silt, as well as fertilizer, pesticides, and other chemicals from agricultural, industrial, and residential runoff. A drainage ditch, its detractors call it. The critics are as right as they are wrong. The Mississippi River, for example, carries so much fertilizer that it annually creates a sixty-five-hundred-square-mile Dead Zone in the Gulf of Mexico.

But at the same time, the Upper Mississippi National Wildlife and Fish Refuge offers one of the largest havens for wildlife in the nation. Established in 1924, the refuge covers 240,000 acres along 261 river miles from Wabasha, MN (250 miles as the crow flies from Lake Itasca) to Rock Island, IL. The refuge, with its forested bluffs jutting up to six hundred feet above the river, hosts 306 bird species, 250 bald eagle nests, 5,000 heron and egret nests, 50 percent of the world's canvasback ducks, 51 mammal species, 42 mussel species, and 111 species of fish. Forty percent of the nation's waterfowl use the Mississippi, including the refuge, as their migration highway in spring and fall.[38]

As for me, I lament the departure of Canada geese each fall and greet their return in the spring as they follow the Mississippi River and its watershed tributaries, collecting from and returning to the east and the west and the north, forming a vast overhead network like the veins of a single, overarching leaf in the sky.

---

Robert and Kathy invited Andy and me to dinner at their rural home outside of Bemidji the evening after we'd surveyed the park with Robert. Robert's kenneled hunting dogs greeted us as we pulled in, the January sun waning in the frigid western sky. We told stories well into the night. Kathy had wrapped up a fundraising position at Bemidji Public Television and was about to embark on a new position fundraising for the Page Education Foundation. Their daughter had made Itasca-themed presents out of popsicle sticks and paper for her uncle Andy and me. Robert, who abounded with information and stories earlier in the day at Itasca, was quieter now in the evening. We learned, though, that his retirement—originally intended just weeks away—was on a short delay so that their daughter could bring her classmates to Lake Itasca in May and be introduced to Daddy's Park.

The house was warm against a cold night.

When Andy and I returned to Itasca, to the Four Seasons cabin complex, under a starlit sky and in biting cold, another occupant had settled in at the four-plex. Probably an ice-fisherman.

We settled in for the night.

Itasca State Park has a long tradition of hosting visitors, whether in grand style or in primitive sites. A mere decade after the park's founding, the Itasca Park Lodge was built—later to be renamed the Douglas Lodge after Minnesota attorney general Wallace B. Douglas, who had fought fiercely to establish the park. The lodge was constructed with rustic downed pine timber from the Itasca forest. It was designed to house the Park Commissioner and more than two dozen guests, "with sixteen rooms and numerous and commodious closets," a 36-by-22-foot "great room" walled with oiled and shellacked "rough logs," and a large stone fireplace.[39]

Since then, of course, additional cabins have been constructed at Itasca, some of them rustic and some more elaborate, but none with the grand elegance of the Douglas Lodge. And the park offers over two hundred campsites for tents, trailers, and RVs.

I am writing at 7 a.m. from the kitchenette of the Four Seasons cabin on January 10, just a single light on above me near the stove where I have heated up some coffee. It's still dark outside, but the day begins to lighten around seven thirty. Andy is still asleep. Outside it is five degrees Fahrenheit, already up from minus five degrees Fahrenheit when we returned to the cabin last night. The Weather Channel announces that the Lake Itasca region will be warmer than normal in the thirty-day forecast.

Our time with Robert having been completed the day before, Andy and I retrieved our snowshoes from the car and headed north along the eastern lobe—or antler—of Lake Itasca. We'd "discover" the headwaters on snowshoes, we figured.

But as an hour or more ticked off and our progress had been slight, we ceded the headwaters discovery back to Schoolcraft and set a new goal. We'd cross Lake Itasca on snowshoes. After all, there were numerous pickup trucks parked on the ice in the middle of the lake while their owners fished from an assortment of huts.

Twenty feet out onto the lake, our shoe prints filled with water. This was not good. We tried another spot, and again our tracks filled with water about twenty feet from the shore. It could have been that the underlying ice was solid and a thin layer of slush had accumulated in the insulating snow cover. After all, those pickup trucks out in the center of the lake. . . . But then, I had read (from Schoolcraft, among others) that Lake Itasca is fed by lake-bottom springs, and perhaps at certain places the spring water weakens the ice cover from underneath. The ice fishermen may have known the safe route onto the ice. We were not ice fishermen.

We abandoned the hike. Our wives and colleagues, via Facebook, thought it a wise choice. Our adult children were delirious with derision. We were too embarrassed to ask Robert. That night, back at the cabin, I Googled the matter. Slush on top of the ice may be normal and entirely safe. Unless there are springs underneath.

The cold followed us back to Dubuque the next day, traveling faster than the river waters. The winter had been mild so far back home, but within a few days of our arrival, a six-inch snowfall pummeled the town. A day after the snowfall I was on the road again, this time with nine Loras College students en route to the other end of the Mississippi, to New Orleans, for a Gulf Coast environmental restoration service trip. The delta is sinking relative to rising sea levels, in part because the channelized and dammed river no longer spreads its replenishing hoard of silt out over the bayou. It is all one connected thing.

My wife and I returned to Lake Itasca six months later, in June. Our family was vacationing at a Minnesota lake two hours away, and one rainy day, we ditched the cabin and headed north to the headwaters.

We won't call it an idyllic day. The mosquitoes bit us ravenously. The headwaters trail was crowded. But we played for a while, as we should have, in the infant Mississippi waters.

Driving home later in the week, we watched the landscape change before us. The North Country's vast flatland forests and eyeball lakes gave way to rolling hills, and then to steeper and steeper rock towers in the Driftless. We had exited one region of mystery and reentered another.

But as unique as these regions are—how utterly unlike some Fly-Over airline passenger's vision of the Midwest—it occurred to me on our drive

Summer visitors like to wade across the Mississippi River at its emergence from Lake Itasca in northern Minnesota.

home how much is connected in some way to this river. We crossed the Minnesota River, which joins the Mississippi at the Twin Cities. Then the Root River, the Zumbro, the Upper Iowa, all flowing to the great river.

The Mississippi River enters the Driftless Area just south of the Twin Cities. If river lengths have a life, the Mississippi leaves its youth behind in the North Country. By the time it reaches the Driftless it has entered its prime, bisecting steep river bluffs. Every drop of water in every Midwest stream (and well beyond) wends a path to the Mississippi. By New Orleans the river has aged but carries with it the weight and power of wisdom.

But here at Itasca begins our Mississippi River, clean, clear, and tumbling forth over a cascade of rounded stones.

And here in the thirty-foot-wide river with its glacially pebbled bottom play the children from various lands, all speaking a common language.

The Ojibwe did not understand the White men's obsession with finding the headwaters. All the river's waters and tributaries played a role in watering the earth.

Jacob Brower echoed the Ojibwe in at least one small way. While he declared once and for all that the Mississippi River began at and flowed from Lake Itasca, he conceded that the waters flowing into Itasca from several upstream lakes and creeks contributed to the Mississippi River as well. He called these upstream sources the "Greater Ultimate Reservoir."[40]

He almost said it, but not quite: *It is all one thing.*

# 12

# Iowa

## *Effigy Mounds in the Driftless Land: The Return of the Old Ones*

We have just started up the steepest slope of the 450-foot climb to the Marching Bears at Effigy Mounds National Monument when David Barland-Liles pauses our group along the trail. We are nine Loras College students, myself as their faculty moderator, and Eric Anglada and Brennan Cussen, a Catholic Worker Farm couple who'd organized this Sacred Lands tour. (David at the time was lead ranger in charge of law enforcement at Effigy Mounds.)

Effigy Mounds in northeast Iowa is a national monument established in 1949 to protect over two hundred Native American burial and ceremonial mounds dating from 500 BC to AD 1300. In succession over time, the earthworks took form as conical, linear, compound, and effigy mounds, the lattermost in the shapes of birds and bears.

The Marching Bears group is the final and most elaborate expression of the mound-building culture here, with ten bear-like effigies in line as if they are traveling south, their feet pointed eastward to the Mississippi River. An eagle effigy corrals them from the rear, and a carrier (passenger) pigeon and peregrine falcon lead them at the front. The largest bear of the group is one hundred feet long; the wingspan of the largest bird is 212 feet.

But we've got a serious climb ahead of us before reaching the Marching Bears, and already some of us are huffing. The brief pause is welcome.

"There are 116,000 human remains waiting to be repatriated to Native Americans," David tells us, many of them from excavated mounds such as the twenty-five-hundred-year-old conical mound behind us near the base of the trail.[1] Just a bit uphill from the conical mound is a linear mound, constructed two thousand years ago. Fortunately, archaeological excavation of mounds ceased here in the early 1970s, but not before the remains of forty-one individuals were unearthed, studied, and filed away at Effigy Mounds' offices. Remains were displayed in the visitors' center until 1973.

"These mounds were built with such precision that they still exist today so that the venerated persons buried here are protected on their ongoing journey," David explains. Archaeologists have been stunned, he says, by how well the mounds were constructed to withstand erosional forces over time. The mounds were built up layer by layer with river soil, clay, and sand, much of it carried up in baskets from the Mississippi River bed 450 feet below the Marching Bears.

But the mounds can't be considered separate from their landscape. "During the Ice Age this place was a refugium," David says with a sweep of his hand, referring to northeast Iowa's portion of the Driftless Area, the region that was largely spared the onslaught of the glaciers, the last one ending a mere twelve thousand years ago. People lived here less than one hundred miles from the edge of the great ice. Tools have been found here dating back 10,000–15,000 years, "since time immemorial," David says, using the phrase that tribal partners[2] have taught him.

The rugged lands of the Driftless Area—or the Paleozoic Plateau, as geologists call it—defy outsiders' impressions of the state. Most know or think they know Iowa as flatlands or, at most, gently rolling hills filled with endless row crops. The late Public Broadcasting Service journalist Bill Moyers once tagged along on RAGBRAI (the *Des Moines Register*'s Annual Great Bicycle Ride Across Iowa), the nationally famous weeklong bike ride across the state, and reported that the scenery of Iowa is "ever changing: sometimes the corn is on the right and sometimes it's on the left!"

But *this* place!

Here in northeast Iowa, limestone bluffs rise 300–450 feet above the Mississippi River. Inland valleys cut sharp V's into the hillsides, downcut by rushing streams in heavy rainfall and snowmelt. Roads curve along creek bottoms and then exploit the gaps between hills to climb back up to the ridge tops. Rock towers poke out from the soil like exposed bones of earth. Sinkholes dot the landscape where bedrock limestone has dissolved and collapsed. Streams and caves run beneath it all. Springs emerge where underground streams meet the slope of the earth.

David relates what the Effigy Mounds' tribal partners have told him about this region: "This place isn't special because the mounds are here. This place is special. That's why our mounds are here."

---

Just south of my home in Dubuque, Iowa, along the Mississippi River and at the southern edge of the Driftless, lies the Mines of Spain State Recreation Area. Opened in 1981, its fourteen hundred acres sit on lands once belonging to the Meskwaki, then mined for lead by Julien Dubuque and the waves of miners who followed him.

Its name was the contrivance of Julien Dubuque, a French Canadian who was the first Euro-American permanent settler to live in what is now Iowa. He arrived in 1788 and secured permission from the Meskwaki to mine for lead in a three-mile inland swath along seven miles of the Mississippi shoreline. But to hedge his bets, he also secured permission to mine the hills from the governor of New Spain in Louisiana during the brief period when Spain claimed sovereignty over most of the land west of the Mississippi. Dubuque promised to call his workings the Mines of Spain.

Dubuque appears to have had comparatively good relations with the Meskwaki. He may have married a member of the tribe. The Meskwaki often worked alongside him. After Dubuque's death in 1810, the Meskwaki buried him next to Chief Peosta on a prominent bluff overlooking the river.

The Meskwaki were forced to move west after the 1832 Black Hawk War, even though they weren't directly involved. Lead miners already actively working across the river in Illinois hurried over to Iowa a year later when the land was legally opened to Euro-Americans.

Chief Peosta and Julien Dubuque rested undisturbed in their graves until 1897, when the townspeople of Dubuque (named for, but not founded by, the French Canadian) decided that Julien deserved a more fitting monument. They dug up the remains of Dubuque and Peosta, and reburied Dubuque within a handsome, limestone-turreted monument at the same location overlooking the confluence of Catfish Creek with the Mississippi, above where the Meskwaki village once lay. The monument looks like a chess castle rook and is a prominent icon for anyone boating on the river due to its perch atop the sheer limestone bluffs.

Peosta's remains were removed and displayed in a local museum for decades. I remember seeing the garish sight when I was a kid. He was finally reburied near, but not next to, Julien Dubuque in 1973.

A castellated monument marks the grave of Julien Dubuque, a French Canadian who mined lead alongside the Meskwaki from 1789 to 1810, just south of present-day Dubuque, Iowa's oldest city. The Meskwaki buried Julien Dubuque and Chief Peosta together at this prominent overlook above the Mississippi River, although Peosta's remains were later separated from Dubuque's.

The Native American Graves Protection and Repatriation Act, or NAGPRA, went into effect on November 16, 1990. On July 16 of the same year, anticipating the passage of the law, then-superintendent of Effigy Mounds Thomas Munson removed the remains of forty-one Native American individuals from "storage" at the monument's headquarters. The remains had

been disinterred by archaeologists who had partially excavated at least thirty of the two hundred mounds. The excavations took place between 1949 and 1971, when the practice ended. However, the remains—the Old Ones, as tribal members call them—had never been repatriated to the tribes. Munson was determined that they not be returned.

With the help of another employee, Munson packed the remains into garbage bags, placed them into two cardboard boxes, and loaded them into the trunk of his car. To cover his tracks, he filled out a form normally used for disposal of government property, such as a microwave or a coffee maker that no longer works. He checked the box on the form that said "Miscellaneous material that does not meet the scope of collections for Effigy Mounds N.M." For disposal, he checked "Abandon."

Munson took the two boxes to his home across the river in Prairie du Chien, WI, and placed them in his garage. Munson retired as superintendent in 1994. The Old Ones sat in his garage for over twenty years until US Park Service special agent David Barland-Liles discovered them in 2012.

David is continually out in front in the hike up to the Marching Bears. He is sometimes by himself slightly ahead of the group, but is usually in conversation with the students and the older adults who rotate up for a chance to talk with him. He stoops occasionally to scoop up even the smallest scrap of litter that desecrates the sacred grounds. A student jokes about the challenging uphill climb. David adds, "Now imagine it carrying baskets of dirt!"

One student brings up the topic of Native-American-themed mascots, saying her high school teams were called the Chiefs. "No one ever gave a thought about it," she says.

"That will change," David responds. "They may not know it yet, but that will change."

Near the top of the trail, before we reach the Marching Bears, David pauses again. He points to a large clearing in the forest, explaining that most of the region would have looked like this at the time of First Contact: an oak savanna, with large oaks spread far and wide and with prairie grasses stretching out beneath them. Native Americans tended these oak savannas and the prairies farther inland by periodic burning. The burned prairie grasses quickly replenished, green and lush, attracting game like bison on the plains and elk and deer in the oak savannas.

David mentions that when Marquette and Joliet became the first Europeans to enter the Upper Mississippi River in 1673—not far from this location—they noted the fires on the river bluffs for miles around.

Without fire, the forest encroaches. "We're in Iowa, everything grows here!" David laughs. But keeping an oak savanna or prairie intact through burning is labor-intensive. The thick forest that encircles us grew in after the tribes had been forcibly removed and the fires were no longer maintained. However, the Park Service has recently been talking with the Ho-Chunk about how to do prescribed burns to recreate more oak savanna in a manner that is still respectful to the sacred grounds.

Discussion shifts slightly to the care of the mounds themselves. Trees must be kept from taking hold on the mounds, as tree roots damage them. But routine mowing would be irreverent. As a result, David says, two Ho-Chunk employees mow the mounds twice a year. National Park Service biologists ensure that invasive species do not take hold. Native plants thrive on the mounds and return each year, as David puts it, "automagically."

Our conversation transitions briefly to environmental issues. David reminds us that indigenous peoples lived here for fifteen thousand years with a relatively soft imprint on the environment. "We will have to learn from them how to live on this land," he says.

We are nearing the Marching Bears. David relates an insight from the tribal partners regarding the burials ahead of us: "We don't own anything. The land owns us. The most important thing we do is to continue on our forever journey."

Sometimes I feel like the Driftless Land possesses me. When I returned from a four-month stay in Ireland several years ago, the first place I went with Dianne to reconnect with the Driftless was Effigy Mounds.

Elsewhere the Driftless calls me as well. The Driftless straddles four Midwestern states—northeast Iowa, southeast Minnesota, northwest Illinois, and southwest and central Wisconsin. Some people call the Driftless the Midwest's thirteenth state. Dianne and I hike its hills, bicycle its roads, and kayak its streams.

But the Iowa Driftless, this narrow northeast band along the Mississippi River, is home. Because I know it more intimately, here in particular, the Driftless speaks to me not just of the landscape but also of the human story layered on it like the duff of oak leaves entering the soil each fall.

On solstices and equinoxes, and on the Celtic quarter-feasts in between, I drive to greet the sunrise over the Mississippi River at the Julien Dubuque grave and monument, keeping vigil with the old miner himself.

Each winter, I take students to a Mines of Spain bluff above the river where surface lead mine pits—four to ten feet deep and up to twenty feet wide—dot the woods like a bomb crater field. Dug by hand in the 1800s, they've lain undisturbed since then while the forest has grew back around them. I feel the presence of the miners in the soft snow that etches the lips of the mine pits.

At White Pine Hollow, an Iowa state forest, pines stake out one slope, oaks another, with rock towers rising in between. Springs erupt from the rocks. A now-deceased friend we used to hike with there said that

Northeast Iowa's Effigy Mounds National Monument preserves and protects over two hundred Native American burial mounds—conical, linear, and bear- and bird-shaped effigies on bluffs 450 feet above the Mississippi River. The mounds were constructed between 500 BC and AD 1300.

my musician-wife crossed the creek on stepping stones as lightly as she touched her piano keys. I remember him each time we go there.

At Backbone State Park, the layers include my parents taking my brothers, sister, and me there once each summer for a rare day away from home. We swam in the artificial lake next to the Civilian Conservation Corps (CCC) boathouse. We rushed over the exposed bedrock on the tall hogback that narrows to a rock tower. Two stream valleys 150 feet below on

either side converge beneath the tower to become the Maquoketa (or Bear) River. Dianne and I took our own kids to Backbone, and we go there now, by ourselves, since our kids have grown to adulthood.

We occasionally paddle the Upper Iowa River, where small, clear waterfalls splash off the bluffs nearly into our kayaks. In Decorah, Dunnings Spring births a full-blown creek from an opening in the rocks as if Moses had tapped the stones. My shoes and socks come off—again!—in the stream once it lands and hurtles toward the Upper Iowa River. It is painfully cold on the ankles and shins, but it must be waded.

In northeast Iowa's Clayton County near the Mississippi River, a battle rages between a local sand excavating company and environmentalists. Near the end of the glacial period an ice dam burst upstream on the Wisconsin River, sending a torrent of lake water and accumulated sands downstream and on into the Mississippi. When the deluge receded, deep sand deposits were left behind in southwest Wisconsin and for miles along Iowa's Mississippi shoreline below the rivers' confluence. The sand deposits are noted for their purity. Mined sand was once merely shipped to line the beaches of artificial lakes and for children's sandboxes, but now its qualities are prized for oil and natural gas fracking in distant states. It provides good jobs for an impoverished county. Environmentalists dislike both its destination use and its ecological and aesthetic impact on the Driftless landscape.

When Jacques Marquette entered the Mississippi River from the confluence with the Wisconsin in 1673, the "mountains" he joyously described on the far side of the river were the bluffs of today's Pikes Peak State Park. In 1805, a year after President Thomas Jefferson famously sent Lewis and Clark to explore the Missouri River, he sent Zebulon Pike to investigate the Upper Mississippi. Along the way, Pike met with Julien Dubuque, who successfully dodged Pike's inquiries about his lead-mining operations. Pike searched unsuccessfully for the headwaters of the Mississippi. He proposed an American fort for the commanding bluff in the park now named after him, but the army decided to build Ft. Crawford at nearby Prairie du Chien instead. In other words, there are a few dings against Pike's Mississippi River exploration, but his name graces one of Iowa's grandest views that sweeps up and down the Mississippi and across to the Wisconsin River from a 450-foot, bluff-walled ledge.

Iowa's Pikes Peak may not swagger like the towering, same-named (for the same Zebulon Pike) Colorado mountain, but chances are you'll have the view to yourself there or strike up a conversation with the only other couple at the overlook watching history float by.

The story at Pikes Peak is as deep as the bedrock and begins with bedrock: at the top, Galena limestone formed in deep quiet seas 450 million years ago; then Decorah limestone and shale, laced with fossils; Platteville limestone, an impervious layer that creates small waterfalls where it outcrops; St. Peter sandstone, formed from a sandy beach; and at the bottom, at river level, Jordan sandstone, 550 million years old.[3]

Here, too, my own memories run deep, of camping not far from the river bluff, including one fretful night when I accidentally left the tent screen open and we were beset by mosquitoes. We finally gave up and drove across the river to Prairie du Chien to find the last available motel room. So much for the intrepid explorers.

In 1909, area enthusiasts launched a decades-long but unsuccessful attempt to create a national park among the undeveloped shorelines and islands of the Upper Mississippi River in northeast Iowa and southwest Wisconsin. But the wheels of such proposals move slowly, and the 1930s construction of the locks and dams altered the natural features of the river and increased commercial traffic, ultimately leading to the rejection of the proposal.[4]

Several of the areas that would have been included are protected in other ways: Iowa's Pikes Peak State Park and Yellow River State Forest, Wisconsin's Wyalusing State Park (at the confluence of the Wisconsin and Mississippi Rivers), and the Upper Mississippi National Wildlife and Fish Refuge. Much of the areas today, on both sides of the river, are designated Important Bird Areas, a modern refugium.

The federal report that rejected that national park proposal advocated instead that Effigy Mounds be named a national monument. The designation took effect in 1949.

---

David is on the move again, with the students and us older adults in tow. He draws our attention to a slight leveling in the slope of the woods that eventually joins our path. This is the military road along which the Ho-Chunk and other southwest Wisconsin tribes were forcibly removed in the years following the 1830 federal Indian Removal Act and the aftermath of the 1832 Black Hawk War. The US military ferried Wisconsin Ho-Chunk and other tribes—who were not even directly associated with the war—across the Mississippi River from Ft. Crawford at Prairie du Chien and marched them up the steep river slopes on the Iowa side at today's Effigy Mounds. They were forced to walk past their ancestral burial mounds en route to their first displacement at Fort Atkinson, Iowa, forty miles away. The path became known as the Trail of Death.

"Tribal partners have asked me to carry tobacco and spread it along the trail in order to comfort the Old Ones," David says, producing a pouch and passing it around to the rest of us. He invites us each to take a pinch and spread it as we walk, wherever we are called to do so. For a while there is no talking among the group as we walk quietly, deliberately, reflectively. I drop pinches from my fingers each time I pass by an old tree along the trail.

Although the theft of the human remains from Effigy Mounds occurred in 1990, the act wasn't fully understood until much later. What first brought Park Service special agent David Barland-Liles to Effigy Mounds was a probe of seventy-eight projects in violation of the National Historic Preservation Act that took place during Superintendent Phyllis Ewing's eleven-year tenure from 1999 to 2010. The infractions included construction of boardwalks on the grounds without consulting tribal partners. The new boardwalks disturbed some mounds and their perimeters, inappropriately looked down on mounds from above, and cut through mature forestland.[5]

Around the time the Ewing investigation was winding down, a tribal representative inquired about the disinterred human remains at Effigy Mounds. In response, Effigy Mound Park Service rangers tracked down a report indicating that remains had been sent to the state archaeologist's office in Iowa City in 1986 for further study, but discovered that no remains had ever been delivered there. Something was definitely amiss.

Effigy Mounds superintendent Jim Nepstad, who had replaced Ewing in 2011, found himself with the unenviable task of telling tribal partners that the Old Ones couldn't be found.

The staff meanwhile consulted with Munson, who returned a box of bones to Effigy Mounds, saying he thought they were animal bones that had been mistakenly left at his home. But when a ranger opened the box back at the offices, he found human remains within, tucked inside a garbage bag. And they didn't account for all forty-one persons whose remains had been removed.

Barland-Liles was called back in again to investigate.

Finally, an employee admitted that she had been ordered by Munson in 1990 to box up the remains. She had placed two boxes in the trunk of Munson's car.

David figured that the tribes had been deceived, stolen from, and injured enough by the national government in regard to these sacred grounds, so he took a new approach. He decided at the outset that the inquiry into the stolen persons would become the first federal investigation in history to be monitored by sovereign indigenous nations. "I shared everything I found," David says.[6] Four tribal representatives occasionally joined in on the investigation as well.[7]

Eventually Munson's fraudulent "disposal" form turned up, leading David back to Munson's house. Munson offered various explanations of what might have happened to the remains, each one of which David shot down. Finally, Munson's wife looked her husband in the eyes and said, "I don't think you're telling this young man the truth."

David produced a "consent to search" form, which Munson's wife signed. She then slid the form in front of her husband and repeatedly tapped where he should sign it. Tom reluctantly signed.

Munson's wife then led David to a detached garage where they quickly found a box. She opened the box and trash bag and found herself looking at the tip of a human femur. "She backed away from it," David recalls.[8]

Barland-Liles took the Old Ones back to Effigy Mounds.

After the evidence had been gathered, David consulted with tribal partners as to what charges should be brought against Munson. He presented about fifty options. They chose that Munson be charged with "theft of government property."

The tribal leaders' choice of charge suggests to me a deep sense of irony. As David says, "Munson did this racist act to circumvent NAGPRA." But if Munson had considered the Old Ones to be nothing more than "government property," then that is what the tribal leaders asked that he be charged with stealing.

David presented the case to the US Attorney's Office in July 2012. After delay tactics were exhausted and appeared futile, Munson finally pled guilty on January 4, 2016.

On July 8, 2016, Munson was sentenced to one year of custody, including house arrest (monitored with ankle bracelet) and weekends in jail; a public apology; a $3,000 fine; $108,905 restitution to Effigy Mounds to pay for all aspects of repatriation, including tribal partner participation; and one hundred hours of community service.

Munson died shortly after completing his sentence.

The healing process began gradually. David accepted a reassignment to Effigy Mounds as lead ranger in January 2018. ("Don't call me 'Chief Ranger,'" he tells our group, rejecting the title that had formerly been applied to his position. The Park Service had done enough usurping of Native American traditions and rights, and it needed to end.) Slowly the tribal partners began to trust the new leadership at Effigy Mounds, with Nepstad, Barland-Liles, and others. They welcomed the hiring of cultural resources program manager Albert LeBeau, a member of the Cheyenne River Sioux Tribe of the Lakota Nation. In 2018, tribal members began expressing to the Park Service their desire to return to Effigy Mounds to openly perform sacred ceremonies.

When the Old Ones were finally repatriated, David was asked to rebury them at Effigy Mounds in a tribal ceremony in fall 2020. The request was double-edged. It was, in part, punitive. A White man was called upon to do the reburial because "it was your people's fault" that the excavations and theft had occurred in the first place. But at the same time, they were honoring David for his work in recovering the Old Ones.

The tribal partners have given David affectionate, or "achievement," names. The Winnebago call him "Carries Our People." To the Dakota he is "Casts Aside" (meaning that he casts aside obfuscations to get to the truth). For the Pottawattamie: "Walks like Talks."

David has been telling this story to the students as we've ascended the trail. He sums up his work as an investigator by saying, "Most of the criminal activity that has occurred here has not been done by looters or ATVs (all-terrain vehicles) or such, but by the National Park Service itself."

He will be taking on new duties soon, he tells us, having just accepted a position as investigator for NAGPRA, a newly created full-time position that will put him in the thick of, among other things, the repatriation of Native Americans buried in boarding school cemeteries.

The mood changes as we emerge from the forest trail onto the hilltop clearing where the Marching Bears wait for us. We are joyous, reverent, amazed. The autumn day is brilliant under blue skies and the first hint of leaf color. The Mississippi River, 450 feet below, glints through a break in the trees. We sense the spirit here.

The confluence of the Wisconsin and Mississippi Rivers is integrated into the Ho-Chunk creation story, David says: "If where you began as a people is also where you are buried, then you are highly buried."

The bears and birds stretch out before us in suspended animation. Four bears, including the two largest, mill about in back, getting into formation. The rear eagle is ready to swoop off the bluff. Six more bears are on the move, head to toe, along the grassy curve of the bluff. The pigeon and falcon are out front at a slight distance, leading the pack, but ready, too, to take flight out over the river.

But where are the bears going? What are the mounds telling us? David relates a story from the oral tradition as it was told to him: "We became humans from the Great Bear. And one night the Great Bear had a dream in which he learned that the people had to separate and find other places in order to prosper." And so the bears and birds—each likely associated with respective clans—departed from this central place. Some marched away along the river, and some flew across it. They spread out in all four directions and became the many tribes.

David invites us to wander quietly among the mounds, reminding us to be respectful of "the work the Old Ones need to do." For a while we circle the mounds in silence and look out over the Mississippi where the eagle is pointed. We discover something about living with mystery, in awe of that which we can't fully explain.

Then David brings us together for one more celebrated story. Some mounds, he explains, once had a ceremonial fire ring, often at the location of the animal's heart. Park Service personnel had just recently discovered that several of the effigy fire rings, plus the tip of a linear mound and the location of a life-giving spring, align perfectly with the Big Dipper on the fall Equinox. The students buzz in wonderment. "It is amazing how precisely, how mathematically these mounds are aligned," David says.

The tribal partners, of course, were aware of the celestial alignments all along, David adds, but were glad that the Park Service had finally figured it out. But they have declined to further discuss the meaning of the alignments. According to David, "They've said, 'We're glad you've noticed, but this is our culture.'"

Perhaps full atonement will never be achieved. But in addition to making things right with tribal partners at Effigy Mounds, I can't help but think that we also atone by learning that the land we walk on is sacred, by learning to listen to its stories. That is, at least, what I hope to teach my students.

The forty-one Old Ones are back at Effigy Mounds, back at the refugium.

Effigy Mounds is the sacred jewel of Iowa's Driftless Region. But the sacred goes on forever in this mystical landscape. The eagle effigy near the rear of the Marching Bears is poised to swoop out over the Mississippi River. From there, no doubt, it will glide past the exposed bluffs at Pikes Peak across from the mouth of the Wisconsin River. It will fly inland over the sweet-flowing Dunnings Spring. It will ride the updrafts along the Upper Iowa River, and then bank a turn above the cornfields that wrap around the contoured hills. In a couple of wing flaps, it will circle over at the Mines of Spain, in Dubuque, where this morning's autumn sunrise is lighting up the red, orange, and brown foliage on the river bluff at the graves of Julien Dubuque and Chief Peosta. I have seen the eagle there. I have seen its white head and tail gliding above my kayak in summer and perched high in the shoreline trees. I have seen it perched on the edge of river ice in winter, waiting for an errant bass to surface.

And I know that the eagle returns each night, or when it is ready, or when it is called, to its sacred place in the soil at Effigy Mounds.

David finishes the story at the edge of the last mound, the peregrine falcon. Right behind him is the bent oak where he first observed, in the company of tribal partners, the Big Dipper blazing itself into the mounds. The students gather around him. He seems to share the gift of the oral tradition. Finishing the Munson story, he tells us that he placed the box containing the recovered Old Ones on his front car seat to return them to Effigy Mounds. "As I pulled away from the Munson house, Tom and his wife were standing on the doorstep looking like the figures in *American Gothic*," he recalls.

I imagine him breathing a sigh of relief as the Mississippi River came into view and he crossed back into Iowa.

"I pressed the play button on my iPod," he continues. "The song playing was 'Testify,' by Rage against the Machine. You know, the one that goes, 'Who controls the past controls the future; Who controls the present controls the past.'" The students grin in recognition.

"So I put my free hand on the Old Ones and said, 'Is this OK?' And they said, 'Turn that shit up!'"

# Conclusion

## *Midwest Bedrock*

I am partial, of course, to limestone, the bedrock of my home in northeast Iowa. It lines the Upper Mississippi River in three-hundred-foot and four-hundred-foot sheer bluffs and rises unexpectedly from the creek valleys in rock towers like the bones of earth. It is the foundation of our older homes, mine and Dianne's included. It is the mysterious maker of caves and the bed of fossils. It is the substratum of the Midwest.

Shale and mudstones speak of a workaday Midwest. Formed from silt, this bedrock foreshadowed prairies and row crops and dirt under the fingernails.

Sandstones are flashier. From clean-sanded sea bottoms, midwestern sandstones were uplifted and sculpted by rain, wind, and rivers. Flares of green and orange blaze through the yellow rock. The Midwest can do that, too.

We do other bedrock as well. We do quartzite, reshaped under heat and pressure into pink and purple angular stone that rises to the sky in jagged defiance.

We do minerals and ores: copper and iron and lead. We have a history of finding treasure in the earth.

And we are deep loamy soils bequeathed to us from our prairies. Here our bedrock lies buried. Sometimes we leave things unspoken.

Dianne and I returned home from Michigan's Upper Peninsula in July 2022, from the last of our Midwest tours. We'd driven, camped, hiked, and bicycled along the Lake Michigan, Huron, and Superior shorelines, the largest freshwater lake system in the world.

We'd traveled through North and South Dakota. Outsiders who think they know the Dakotas may not have anticipated a preponderance of glacial lakes, as well as escarpments that rise quickly above the plains. And not everyone knows how the plains' prairie grasses blow in the wind like ocean waves, and how a hill overlooking the prairie is an experience of the Holy.

We'd driven through eastern Kansas and northern Nebraska through lands defying Midwest stereotypes. Kansas's Flint Hills roll across a prairie landscape, too rugged with surface stone to plow. The Niobrara River slices through northern Nebraska, a swift-moving stream downcutting to a bedrock floor, exposing fossils and drawing to it numerous waterfalls.

Southern Ohio's Hocking Hills is a more visited place, but not well known outside the region. Enormous curved rock shelters and box canyons and waterfalls turn the landscape inside out. A cemetery in the middle of the woods in southern Indiana's Hoosier National Forest reminded us that this vast timberland had once been farmland cleared from forest, abandoned, and then reforested. Remember that you are forest, and to forest you will return.

We'd ridden on Missouri's Katy Trail, flanking the Missouri River and exposing in layers the start and finish of the Lewis and Clark expedition, the beginning and end of the MKT Railroad, and the end and rebirth of towns along the path. In Illinois we climbed the one-hundred-foot-tall Monks Mound, the largest pre-Columbian earthen structure of North America, marking the center of the abandoned Native American city of Cahokia.

Minnesota and Wisconsin were our old haunts, revisited. Living alongside the Mississippi River, how could we resist exploring Lake Itasca, the Minnesota source of the river that is in turn the source of our home? We regularly bicycle, camp, and explore Wisconsin, from its northern uplands to its eastern Great Lakes shoreline, and the rolling hills of the unglaciated Driftless region in between. The Ice Age Trail takes in all these landscapes and more, following the final moraine of the Wisconsinan glacier.

Dianne and I are Iowans, but more specifically, northeast Iowans. Iowa harbors a corner of the Driftless region, too, a surprise to those who thought they knew our state as a level plain. And the apex of Iowa's Driftless is Effigy Mounds National Monument where Woodland Period Native Americans built hundreds of burial and ceremonial mounds 450 feet above the Mississippi River, some in the shapes of bears marching along the bluff and raptors marshaling them along. This is the center of the world.

Like Thoreau, we've traveled widely in our home. This tour of the Midwest has brought surprises in lands we thought we knew, and deepened our understanding of the familiar.

Midwestern landscapes may be more varied than outsiders realize. But in the end even the outsiders get one thing right: this place is the bedrock.

# Notes

## Introduction

1. "The Two Hundred Largest Cities," World Population Review.
2. "State-by-State Visualizations of Key Demographic Trends," US Census Bureau.
3. Hollandbeck, "In a Word," *Saturday Evening Post.*

## 1. Wisconsin

1. "Dalles of the St. Croix River," Wisconsin Department of Natural Resources.
2. Dott and Attig, *Roadside Geology of Wisconsin,* 92–93.
3. Parts of The Wisconsin Ice Age and Chippewa Moraine sections are adapted from the chapter "The Ice Age" in the author's earlier book, *Driftless Land.*
4. Gannon, "Why Do Ice Ages Happen?"
5. Dott and Attig, *Roadside Geology,* 36–37.
6. Wisconsin Department of Natural Resources, *Ecological Landscapes,* vii.
7. Dott and Attig, *Roadside Geology,* 21.
8. Ibid., 203–205.
9. Wollmer, interview.
10. Sheldrake, *Spaces for the Sacred,* 7.
11. Dott and Attig, *Roadside Geology,* 215–222.
12. Shaffer, interview.
13. "Baraboo Hills," Wisconsin Geological and Nature History Survey.
14. *Ice Age Trail Guidebook,* 218.
15. Ibid.
16. Koch, "The Fragrance of My Woods," 154.

## 2. Michigan

1. "Glacial History & the Development of Lake Superior," Lakehead Region Conservation Authority.
2. "Sleeping Bear Dunes National Lakeshore History," Oh.Ranger.
3. "Great Lakes Essential Resources: Shipwrecks," University at Buffalo, University Libraries.
4. The island, straits, and bridge are spelled 'Mackinac.' The city on the Lower Peninsula side of the bridge is spelled 'Mackinaw.' Regardless of the spelling, all are pronounced "Mack-in-naw."

5. Boynton, *Fishers of Men*, 15.
6. Ibid.
7. Janzen, "Delaware Mine Dive, 2012."
8. "The Ore Docks in Marquette," Travel Marquette.
9. "First Discovery of Iron Ore," Ironoreheritage.
10. "Iron Mining: Where and Why?," Project Geo, Michigan State University.
11. Krause, *Making of a Mining District*, 32.
12. Ibid., 39.
13. Ibid., 122.
14. Ibid., 224
15. Ibid., 210–212.
16. Ibid., 241.
17. "Coronavirus Has Spared the Keweenaw Peninsula," *Bridge Michigan*.
18. Rooks, interview.
19. Ibid.

## 3. Ohio

1. Camp, *Roadside Geology of Ohio*, 259.
2. "Brief History of Hocking Hills," Ohio Memory.
3. Camp, *Roadside Geology of Ohio*, 260.
4. Ibid., 263.
5. "Ash Cave," Hockinghills.
6. LaRue, "Camp Sherman Contributes," US World War One Centennial Commission.
7. Hancock, *Guide to the Hopewell Ceremonial Earthworks*.
8. "Hope Furnace Ruins," Ohio Department of Natural Resources.
9. "Underground Railroad: Vesuvius Forest," US Department of Agriculture: Forest Service.
10. Jaeger, "Rise and Fall of Industry," 21–22.
11. Camp, *Roadside Geology of Ohio*, 206.
12. Meyers, Walker, and Vollmer, *Carrying Coal to Columbus*, 16.
13. Camp, *Roadside Geology of Ohio*, 269.
14. Meyers, Walker, and Vollmer, *Carrying Coal to Columbus*, 22.
15. Kruse Daniels, interview.
16. Camp, *Roadside Geology of Ohio*, 269.
17. Meyers, Walker, and Vollmer, *Carrying Coal to Columbus*, 38–39.
18. Ibid., 133.
19. Ibid., 136–137.
20. Ibid., 11.
21. Ibid., 142.
22. Kruse Daniels, interview.
23. Kruse Daniels et al., "Legacy of Appalachian Ohio Coal," 59–60.
24. Kruse Daniels, interview.
25. Kruse Daniels et al., "Legacy of Appalachian Ohio Coal," 59.
26. "State of Ohio Invests $2 Million," *Highland County Press*.
27. Mallom, "Ideology from the Earth," 60.

## 4. Indiana

1. Smith and Klabacka, *Archaeological Investigations.*
2. Sieber and Munson, *Looking at History,* 12–13.
3. Dwyer, "It Just Broke Him."
4. Gilbert, interview.
5. Sieber and Munson, *Looking at History,* 21–23.
6. Sieber and Munson, *Looking at History,* 86.
7. Powell, "Indiana Limestone."
8. Ibid.
9. "Orangeville Rise, Indiana," Nature Conservancy.
10. Sieber and Munson, *Looking at History,* 54.
11. Johnson, interview.
12. "The Greatest Good," US Forest Service.
13. Trask, *Black Hawk,* 286.
14. Wilkie, *Dubuque on the Mississippi,* 125.
15. Walker, "Lick Creek Settlement Holds Piece of Black History in Indiana."
16. Ibid.
17. "Buffalo Trace Trail," Indiana Historic Pathways.
18. Ibid.

## 5. Kansas

1. Brown, interview.
2. Least Heat-Moon, *PrairyErth,* 77.
3. Patterson, interview.
4. Ibid.
5. "Last Stand of the Tallgrass Prairie," National Park Service.
6. Ibid.
7. Holm, "Horizontal Grandeur," 44.
8. Least Heat-Moon, *PrairyErth,* 55–56.
9. "Last Stand of the Tallgrass Prairie."
10. "Intersection of Cultures," KawMission.
11. Santa Fe Historic Trail.
12. Brake, *On Two Continents,* 121.
13. Ibid., 122.
14. Stahl, *1865 Santa Fe Trail Diary of Frank Stahl.*
15. Editorial, *Junction City Union,* 1867, qtd. in Bieber, "Some Aspects of the Santa Fe Trail 1848–1880," 162–163.
16. Mandel, "Claussen Archaeological Site."
17. Least Heat-Moon, *PrairyErth,* 582.
18. "Cultural History," Kawnation.
19. Tanner, "Kanza People Returning to Sacred Land Near Council Grove."
20. Hoy, "Stephen F. Jones and the Spring Hill Ranch."
21. "Last Stand of the Tallgrass Prairie."
22. "Chase County, Kansas Population 2021," World Population Review.
23. "Last Stand of the Tallgrass Prairie."
24. "Chase County, Kansas Population 2021."
25. "Chase County, Kansas," 2017 Census of Agriculture County Profile.

26. Conard, *Tallgrass Prairie National Preserve Legislative History, 1920–1996*, 4.
27. Ibid., 26.
28. Ibid., 8.
29. Ibid., 27.
30. Ibid., 28.
31. Ibid., 31, 46, 48.
32. Beam, "Beef Production in the Flint Hills," 103–104.

## 6. Nebraska

1. Manning, *Grassland*, 217.
2. Johnsgard, *The Niobrara*, 5–7, 15.
3. Ibid., 23.
4. Both quoted in Jones, *The Last Prairie*, 78.
5. "Niobrara Valley Preserve," Nature Conservancy.
6. "David Sand's Niobrara Timeline," Friends of the Niobrara.
7. "Niobrara Valley Preserve."
8. "David Sand's Niobrara Timeline."
9. Ibid.
10. Hefner, interview.
11. "Niobrara Valley Preserve."
12. Farrar, "The Niobrara as a Scenic River," 141.
13. Roeder, "Niobrara National Scenic River," 4.
14. "David Sand's Niobrara Timeline."
15. Farrar, "The Niobrara as a Scenic River," 142.
16. Roeder, "Niobrara National Scenic River," 36.
17. Farrar, "The Niobrara as a Scenic River," 144–145.
18. Duggan, "Federal Judges Back Park Service."
19. "Niobrara National Scenic River," National Wild and Scenic Rivers System.
20. Ibid.
21. "From Waterhole to Rhino Barn," University of Nebraska State Museum.
22. Ibid.
23. Gambino, "Evolution World Tour."
24. Ibid.
25. Johnsgard, *The Niobrara*, 27.
26. Ibid., 30–31.
27. "Fort Niobrara National Wildlife Refuge," US Fish and Wildlife Service.
28. Johnsgard, *The Niobrara*, 33.
29. Sandoz, *Old Jules*, 268.
30. "Cherry County, Nebraska." Census reporter.
31. Ibid.
32. *Economic and Social Values, iii.*
33. Sprenger, interview.
34. "Valentine National Wildlife Refuge," US Fish and Wildlife Service.
35. "Fort Niobrara National Wildlife Refuge."
36. Sprenger, interview.
37. "Volunteers Will Drive Longhorn Cattle 180 Miles to Fort Robinson."
38. "Bridges of the Niobrara," National Park Service.

## 7. Missouri

1. All Lewis & Clark *Journal* references are from Lewis, Meriwether, and William Clark, "The Journals of Lewis and Clark: 1804–1806," Project Gutenberg e-book.

2. "Mississippi River Facts," National Park Service; and Bluemle, "North Dakota Geological Survey Notes," ND Geological.

3. Fremling, *Immortal River*, 12, 15, 363.

4. "Missouri River," Encyclopedia Britannica; and "A Comparison of the Missouri and Mississippi Rivers," ArcGIS Online.

5. "Mississippi River Facts."

6. Bluemle, "North Dakota Geological Survey Notes."

7. While many refer to Meriwether Lewis and William Clark as the cocaptains of the Corps of Discovery, only Lewis officially held this title. After President Thomas Jefferson commissioned Lewis to organize the expedition, Lewis offered his close friend Clark the role of cocaptain. Just days before the Corps of Discovery set sail, correspondence from the War Department reached Lewis that Clark had been denied the rank of captain and would be first lieutenant instead. A disappointed Lewis conveyed the information to an equally disappointed Clark, and the two agreed that they would not inform the crew of this decision but would refer to and work with each other as cocaptains on the expedition.

8. The expedition's keelboat (a barge-like, partially covered vessel with a short, steep keel and a flat bottom) was fifty-five feet long with twenty-two oar stations and a thirty-two-foot sail mast. The "red pirogue" (a long, canoe-shaped vessel) was forty-one feet long, with seven oars and a short sail mast. The "white pirogue" was slightly smaller.

9. Rogers, *Lewis and Clark in Missouri*, 30–31.

10. Ibid., 29.

11. "History of St. Charles County," St. Charles County Historical Society.

12. Hessong, "History of the M-K-T Railroad."

13. Hofsommer, "Missouri-Kansas-Texas Railroad," Texas State Historical Association.

14. Hansen, "Katy Trail Landowners."

15. Ibid.

16. Brent, "Thirteen Landowners along the Katy Trail," Missouri Bicycle and Pedestrian Federation.

17. Hansen, "Katy Trail Landowners."

18. Hill and Splinter, interview.

19. "Deutschheim State Historic Site," Missouri Department of Natural Resources.

20. Lensing, interview.

21. "Events Recognizing Black History Month," Visit Hermann.

22. "Our History," Lincoln University.

23. Parks, "York Explored the West with Lewis and Clark."

24. Ibid.

25. Barker, "Providence the Town."

26. Eller and Farren-Eller, interview.

27. Landolt, interview.

## 8. Illinois

1. Koch, "Mesmerizing Mounds." Earlier versions of the author's works referenced in this chapter appeared in the *[Dubuque] Telegraph Herald.*
2. Koch, "Palisades a Treasure."
3. Koch, "A Landing Spot."
4. Koch, "Remembering Black Hawk."
5. Marquette, *Father Marquette's Journal,* 25.
6. Ibid.
7. Ibid., 39.
8. Iseminger, *Cahokia Mounds,* 15.
9. Mink, *Cahokia,* 9–10.
10. Ibid., 13.
11. Ibid., 5.
12. Tyrrell, "As the River Rises."
13. Mink, *Cahokia,* 50.
14. Ibid., 24.
15. Ibid., 31.
16. Anderson-Bricker, interview.
17. Iseminger, *Cahokia Mounds,* 149.
18. Mink, *Cahokia,* 66.
19. Tyrrell, "As the River Rises."
20. Mink, *Cahokia,* 68.
21. Iseminger, *Cahokia Mounds,* 36.
22. Belknap, interview.
23. Anderson-Bricker, interview.
24. Ibid.

## 9. North Dakota

1. "Flood Protection Facts," City of Grand Forks, ND.
2. "Soil Survey of Pembina County, North Dakota," USDA, 2.
3. "MCI to Close Bus Plant in North Dakota," *Bus and Motor Club News.*
4. "The Métis," Pembina State Museum.
5. "SRMSC Construction/Engineering," RSL3 Missile Site Tours.
6. Ibid.
7. "Shutdown of ABM Facility," *The Dispatch.*
8. "Phase-Out / Abandonment of the SRMSC," SRMSC.
9. Finney, "Safeguard ABM System to Shut Down."
10. "Phase-Out / Abandonment of the SRMSC."
11. Finney, "Safeguard ABM System."
12. Ibid.
13. "Unemployment Rate in Cavalier County, ND," FRED Economic Data.
14. Larson, *American Scientist*; and "By the Numbers," North Dakota Water Resources.
15. Breakey, "Hell in High Water."
16. Larson, *American Scientist.*
17. "By the Numbers."
18. Graue, interview.

19. Rooks, "This Date in Native History."
20. "Passion and Persistence Pays Off for Dakota Elder," US Fish and Wildlife Serve.
21. "Fort Totten State Historic Site," Signboards.
22. Ibid.

## 10. South Dakota

1. Norris, *Dakota*, 7.
2. Ibid.
3. Daum, *Prairie in Her Eyes*, 114–115.
4. Norris, *Dakota*, 41.
5. Hillenbrand, OSB, interview.
6. Treinen, interview.
7. Norris, *Dakota*, 1–3.
8 Tvedten, "The View from a Monastery," 312.
9. South Dakota State University Museum of Agriculture.
10. Wahl, interview.
11. *Historic Homes of Watertown, South Dakota*, Codington County Historical Society.
12. Kant, "Historical Geography of Lake Kampeska," South Dakota State University Department of Geography.
13. "Yankton Sioux Treaty Monument," National Park Service.
14. Henning and Schnepf, *Blood Run: The "Silent City"*; and Good Earth State Park Visitor Center.

## 11. Minnesota

1. Robert Chance served as Itasca State Park (Minnesota) supervisor from 2012 to 2019 and began his service with the Minnesota DNR in 1977. He retired six months after the date of this interview. Special thanks as well to Connie Cox, lead interpretive naturalist at Itasca State Park, who gave me a primer on all things Itasca.
2. River miles are notoriously hard to measure. Calvin Fremling in *Immortal River: The Upper Mississippi in Ancient and Modern Times* puts the length at 2,301 miles. A signpost at the headwaters states 2,551. I'll split the difference at 2,400 miles.
3. Fremling, *Immortal River*, 16.
4. Ibid., 12.
5. Martin, "Extinct Bison in Minnesota."
6. "History of the Ojibwe People."
7. "Exploring the Mississippi Headwaters," Read the Plaque.
8. Tyrell, *David Thompson, Explorer.*
9. Pike passed through the Driftless on his way north and visited Julien Dubuque to gather information about the lead mines in the region that would later become my hometown.
10. Pike, *Exploratory Travels through the Western Territories of North America*, 75.
11. *Search for the Great River's Source*, 6.
12. Mason, *Schoolcraft's Expedition to Lake Itasca*, xii–xiv.
13. Davis, *Father of Waters*, 15.
14. Schoolcraft uses the name Lac Le Biche, rather than *Omushkos* or Elk Lake, in his memoir. Thus, I use the name here.

15. Schoolcraft, *Schoolcraft's Narrative of an Expedition*, 33.
16. Ibid.
17. Ibid., 36.
18. Ibid.
19. *Search for the Great River's Source*, 6.
20. Severin, *Explorers of the Mississippi*, 282.
21. Glazier, *Headwaters of the Mississippi*, 339.
22. Severin, *Explorers of the Mississippi*, 283–288.
23. Brower, *Report of the Commissioner*, 18.
24. Itasca State Park Information Center.
25. Chance, interview.
26. Steck, "Logging Boom and the Birthplace of Minnesota."
27. Enger, "History of Timbering in Minnesota," Minnesota Public Radio.
28. Ibid.
29. Cox, *Mary Gibbs*, 10–12.
30. Ibid., 23.
31. Sikes, Tester, and Thoma, *Mammals of Itasca State Park*, 5.
32. Ibid., 28.
33. Ibid., 4.
34. *Sentinel Lake Assessment Report*, 1.
35. Ibid., 3.
36. Ibid., 12, 32.
37. Ibid., 1.
38. "Upper Mississippi River National Wildlife & Fish Refuge," US Fish and Wildlife Service.
39. Cox, *Douglas Lodge*, 7–8.
40. Brower, *Report of the Commissioner*, 18.

## 12. Iowa

1. Barland-Liles, interview.
2. The tribal partners of Effigy Mounds National Monument are: Iowa Tribe of Kansas and Nebraska; Iowa Tribe of Oklahoma; Otoe-Missouria Tribe of Oklahoma; Ho-Chunk Nation of Wisconsin; Winnebago Tribe of Nebraska; Upper Sioux Community of Minnesota; Shakopee Mdewakanton Sioux Community in the State of Minnesota; Lower Sioux Indian Community of Mdewakanton Sioux Indians of Minnesota; Prairie Island Indian Community in the State of Minnesota; Sac and Fox Tribe of the Mississippi in Iowa; Sac and Fox Nation of Missouri in Kansas and Nebraska; Sac and Fox Nation of Oklahoma; Crow Creek Sioux of South Dakota; Omaha Tribe of Nebraska; Santee Sioux of Nebraska; Standing Rock Sioux of North Dakota; Yankton Sioux of South Dakota; Sisseton Wahpeton Oyate; Flandreau Santee Sioux Tribe; and Ponca Tribe of Nebraska.
3. "Geology of Iowa State Parks," Iowa Geological Survey.
4. Wilson, "Park That Wasn't," 41–43.
5. Smith, "Righting Wrongs at Effigy Mounds," 13–14.
6. Barland-Liles, interview.
7. Smith, "Righting Wrongs at Effigy Mounds," 16.
8. "Bears, Birds, and Bones," *This Is Criminal*, podcast #72.

# Bibliography


Anderson-Bricker, Kristin, PhD. Interview by author, November 15, 2021.

"Ash Cave." Hockinghills. Accessed October 14, 2021. www.hockinghills.com/ash_cave.html.

"Baraboo Hills." Wisconsin Geological and Nature History Survey. University of Wisconsin-Madison. Accessed December 1, 2020. http://wgnhs.wisc.edu/wisconsin-geology/major-landscape-features/baraboo-hills/.

Barker, Jacob. "Providence the Town: History, Now for Sale." *Columbia Missourian.* August 9, 2009, updated June 12, 2015. https://www.columbiamissourian.com/news/providence-the-town-history-now-for-sale/article_3a42acac-4a16-55e2-a07a-da9006998od8.html.

Barland-Liles, David. Interview by author, October 18, 2021.

Beam, Mike. "Beef Production in the Flint Hills." *Symphony in the Flint Hills (KS) Field Journal,* 2010. https://newprairiepress.org/sfh/2010/nature/5.

"Bears, Birds, and Bones." *This Is Criminal,* podcast #72, August 4, 2017. https://thisiscriminal.com/episode-72-bears-birds-and-bones-7-3-2017/.

Belknap, Lori, Cahokia Mounds State Historic Site Superintendent. Interview by author, September 23, 2021.

Bieber, Ralph P. "Some Aspects of the Santa Fe Trail 1848–1880." *Missouri Historical Review* 18 (January 1924): 158–166.

Bluemle, John P. "North Dakota Geological Survey Notes #7: The Missouri River." North Dakota Geological Survey. Accessed November 19, 2021. https://www.dmr.nd.gov/ndgs/NDNotes/ndn7_h.htm.

Boynton, James, SJ. *Fishers of Men: The Jesuit Mission at Mackinac 1670–1765.* Mackinac Island, MI: St. Anne's Church, 1996.

Brake, Hezekiah. *On Two Continents: A Long Life's Experience.* Topeka, KS: Crane and Company, 1896.

Breakey, Sharlene. "Hell in High Water: The Story of Devil's Lake, North Dakota." August 24, 2018. https://modernfarmer.com/2018/08/hell-in-high-water-the-story-of-devils-lake-north-dakota/.

Brent, Hugh. "Thirteen Landowners along the Katy Trail Awarded $410,000." Missouri Bicycle and Pedestrian Federation. Accessed November 19, 2021. https://mobikefed.org/2002/12/13-landowners-along-the-katy-trail-awarded-410000.php.

"Bridges of the Niobrara." National Park Service. Accessed August 26, 2021. https://www.nps.gov/articles/000/bridges-of-the-niobrara.htm.

"A Brief History of Hocking Hills State Park." Ohio Memory. Accessed October 14, 2021. http://ohiomemory.ohiohistory.org/archives/2789#:~:text=Over%20330%20million%20years%20ago,what%20is%20now%20Newark%2C%20Ohio.

Brower, Jacob. *The Report of the Commissioner of the Itasca State Park*. Minneapolis: Harrison and Smith, 1893.

Brown, Heather. Interview by author, July 23, 2021.

"Buffalo Trace Trail," Indiana Historic Pathways. Accessed September 29, 2021. http://buffalotrace.indianashistoricpathways.org/.

"By the Numbers: Devils Lake." North Dakota Water Resources. Accessed August 25, 2022. https://www.swc.nd.gov/pdfs/dl_fact_sheet.pdf.

Camp, Mark J. *The Roadside Geology of Ohio*. Missoula, MO: Mountain Camp, 2006.

Chance, Robert. Interview by author, January 9, 2019.

"Chase County, Kansas." 2017 Census of Agriculture County Profile. Accessed September 9, 2021. http://www.nass.usda.gov/Publications/AgCensus/2017/Online_Resources/County_Profiles/Kansas/cp20017.pdf.

"Chase County, Kansas Population 2021." World Population Review. Accessed September 9, 2021. http://worldpopulationreview.com/us-counties/ks/chase-county-population.

"Cherry County, Nebraska." Censusreporter. Accessed August 26, 2021. https://www.google.com/search?q=us+census+bureau+population+cherry+county+nebraska&rlz=1C1GCEU_enUS903US903&oq=US+Census+bureau+population+Cherry+County+Ne&aqs=chrome.1.69i57j33i160l3j33i299.17455j0j15&sourceid=chrome&ie=UTF-8.

"A Comparison of the Missouri and Mississippi Rivers." Arcgis. Accessed November 19, 2021. https://www.arcgis.com/home/item.html?id=2b652a08900f4211a373b587668fe703#overview.

Conard, Rebecca. *Tallgrass Prairie National Preserve Legislative History, 1920–1996*. Omaha, NE: National Park Service, 1998.

"Coronavirus Has Spared the Keweenaw Peninsula but May Kill Summer Tourism." *Bridge Michigan*. May 18, 2020. https://www.bridgemi.com/business-watch/coronavirus-has-spared-keweenaw-peninsula-may-kill-summer-tourism.

Cox, Connie. *Douglas Lodge: Minnesota's Own Resort*. Park Rapids, MN: Itasca State Park, n.d.

———. *Mary Gibbs: A Shining Light for Itasca*. Park Rapids, MN: Itasca State Park, n.d.

"Cultural History." Kawnation. Accessed September 9, 2021. http://kawnation.com/?page_id=4463.

"Dalles of the St. Croix River." Wisconsin Department of Natural Resources. Accessed July 29, 2020. http://dnr.wi.gov/topic/Lands/naturalareas/index.asp?SNA=164.

Daum, Ann. *The Prairie in Her Eyes*. Minneapolis: Milkweed, 2001.

"David Sand's Niobrara Timeline." Friends of the Niobrara. Accessed August 26, 2021. https://friendsoftheniobrara.org/wp-content/uploads/2017/04/Sands-Niobrara-Timeline.pdf.

Davis, Norah Deakin. *The Father of Waters: A Mississippi River Chronicle*. San Francisco: Sierra Club Books, 1982.

"Deutschheim State Historic Site." Missouri Department of Natural Resources. 2021.

Dott, Robert H., Jr., and John W. Attig. *Roadside Geology of Wisconsin*. Missoula, MT: Mountain Press, 2004.

Duggan, Joe. "Federal Judges Back Park Service in Dispute over Protected Land along Niobrara River." *Omaha World-Herald*. May 7, 2018. https://m.norfolkdailynews.com/news/federal-judges-back-park-service-in-dispute-over-protected-land-along-niobrara-river/article_f306f29e-51fc-11e8-8387-43bd088f30f4.html.

Dwyer, Kayla. "'It Just Broke Him': 500 Indiana Families Forced from Homes in 1940 for Weapons Testing." *Indianapolis Star*. December 9, 2021. https://www.indystar.com/story/news/history/retroindy/2021/12/09/indiana-world-war-ii-jefferson-proving-ground-displaced-500-indiana-families/8767390002/.

*Economic and Social Values of Recreational Floating on the Niobrara National Scenic River*. Omaha: University of Nebraska-Omaha College of Business, 2009.

Eller, Eric, and Neely Farren-Eller. Interview by author, September 30, 2021.

Enger, Lief. "A History of Timbering in Minnesota." Minnesota Public Radio. November 16, 1998. http://news.minnesota.publicradio.org/features/199811/16_engerl_history-m/.

"Events Recognizing Black History Month." Visit Hermann. Accessed November 22, 2021. https://visithermann.com/explore-abolitionist/.

"Exploring the Mississippi Headwaters." Read the Plaque. Accessed April 6, 2020. https://readtheplaque.com/plaque/exploring-the-mississippi-headwaters.

Farrar, Jon. "The Niobrara as a Scenic River." In *The Niobrara: A River Running through Time*, edited by Paul A. Johnsgard, 138–151. Lincoln: University of Nebraska Press, 2007.

Finney, John W. "Safeguard ABM System to Shut Down." *New York Times*, November 25, 1975. https://www.nytimes.com/1975/11/25/archives/safeguard-abm-system-to-shut-down-5-billion-spent-in-6-years-since.html.

"First Discovery of Iron Ore in the Lake Superior Region." Ironoreheritage. Accessed August 25, 2022. http://ironoreheritage.com/2017/wp-content/uploads/2017/04/First-Discovery-of-Iron-Ore-in-the-Lake-Superior-Region-08_1-.pdf.

"Flood Protection Facts." City of Grand Forks, ND. Accessed August 25, 2022. https://www.grandforksgov.com/government/city-departments/engineering/flood-control/flood-protection-facts.

"Fort Niobrara National Wildlife Refuge." US Fish and Wildlife Service. Accessed August 26, 2021. https://www.fws.gov/refuge/fort_niobrara/.

"Fort Totten State Historic Site." Signboards. June 25, 2022.

Fremling, Calvin R. *Immortal River: The Upper Mississippi in Ancient and Modern Times*. Madison: University of Wisconsin Press, 2005.

"From Waterhole to Rhino Barn." University of Nebraska State Museum: Ashfall Beds. Accessed August 26, 2021. https://ashfall.unl.edu/about-ashfall/waterhole-to-rhino-barn.html.

Gambino, Megan. "Evolution World Tour: Ashfall Fossil Beds." *Smithsonian Magazine*. January 2012. https://www.smithsonianmag.com/arts-culture/evolution-world-tour-ashfall-fossil-beds-nebraska-6171451/.

Gannon, Megan. "Why Do Ice Ages Happen?" *Live Science*. September 1, 2019. http://www.livescience.com/what-causes-ice-ages.html.

"Geology of Iowa State Parks: Pikes Peak." Iowa Geological Survey. Accessed November 3, 2021. https://iowageologicalsurvey.org/iowa-state-parks/geology-of-iowas-state-parks-pikes-peak/.

Gilbert, Mia. Interview by author, June 15, 2021.

"Glacial History and the Development of Lake Superior." Lakehead Region Conservation Authority. Accessed August 25, 2022. https://lakeheadca.com/events-education/geology/glacial-lakes-history-1.

Glazier, Captain Willard. *Headwaters of the Mississippi*. Chicago: Rand, McNally, 1893.

Good Earth State Park Visitor Center, June 30, 2022.

Graue, Colleen. Interview by author, August 17, 2022.

"The Greatest Good: Pinchot and Utilitarianism." United States Forest Service. Accessed September 29, 2021. https://www.fs.usda.gov/greatestgood/press/mediakit/facts/pinchot.shtml.

"Great Lakes Essential Resources: Shipwrecks." University at Buffalo, University Libraries. June 14, 2023. https://research.lib.buffalo.edu/greatlakes/shipwrecks.

Hancock, John E. *Guide to the Hopewell Ceremonial Earthworks*. World Heritage Ohio, 2020.

Hansen, Rose. "Katy Trail Landowners: Then and Now." *Missouri Life*. December 15, 2019. https://missourilife.com/katy-trail-landowners-then-and-now/.

Hefner, Amanda. Interview by author, July 27, 2021.

Henning, Dale R., and Gerald F. Schnepf. *Blood Run: The "Silent City."* Des Moines, IA: Iowan Books, 2014.

Hessong, Athena. "History of the Missouri-Kansas-Texas AKA Katy Railroad." *Texas Hill Country News*. December 8, 2017. https://texashillcountry.com/history-katy-railroad/.

Hill, Darwin, and Penny Splinter. Interview by author, November 14, 2021.

Hillenbrand, Thomas, OSB. Interview by author, August 8, 2022.

*Historic Homes of Watertown, South Dakota*. Codington County Historical Society. 2009. https://www.flipsnack.com/watertowncvb/historicalhomes.html.

"History of St. Charles County." St. Charles County Historical Society. Accessed November 19, 2021. https://www.scchs.org/cpage.php?pt=24.

"The History of the Ojibwe People." An excerpt from "The Land of the Ojbwe," Minnesota Historical Society. 1973.

Hofsommer, Donovan. "Missouri-Kansas-Texas Railroad." Texas State Historical Association. Accessed November 19, 2021. https://www.tshaonline.org/handbook/entries/missouri-kansas-texas-railroad.

Hollandbeck, Andy. "In a Word: The Surprising Story behind Every State's Name." *Saturday Evening Post*. October 3, 2019. Accessed December 9, 2021. http://www.saturdayeveningpost.com/2019/10/in-a-word-the-secret-meaning-behind-all-50-state-names/.

Holm, Bill. "Horizontal Grandeur." In *Inheriting the Land: Contemporary Voices from the Midwest*, edited by Mark Vinz and Thom Tammaro, 41–45. Minneapolis: University of Minnesota Press, 1993.

"Hope Furnace Ruins." Ohio Department of Natural Resources. Accessed October 14, 2021. http://ohiodnr.gov/wps/portal/gov/odnr/go-and-do/plan-a-visit/find-a-property/hope-furnace-ruins.

Hoy, Jim. "Stephen F. Jones and the Spring Hill Ranch." *Marion County Record*. January 18, 2006. http://marionrecord.com/olddirect/stephen_f_jones_and_the_spring_hill_ranch+m118jones+53746570686 56e20462e204a6f6e657320616e64207468652053707269 6e672048696c6c2052616e6368.

*Ice Age Trail Guidebook*. Ice Age Trail Alliance. Versa, 2017.

"An Intersection of Cultures." KawMission. Accessed September 9, 2021. http://www.kawmission.org/places/kawmission/theplacegrove.htm.
"Iron Mining: Where and Why?" Project Geo, Michigan State University. Accessed August 25, 2022. https://project.geo.msu.edu/geogmich/iron.html.
Iseminger, William. *Cahokia Mounds: America's First City*. Charleston, SC: History Press, 2010.
Itasca State Park Information Center. June 29, 2016.
Jaeger, John. "The Rise and Fall of Industry at Ohio Hanging Rock." *Arc of Appalachia Newsletter*. Winter 2017.
Janzen, John. "Delaware Mine Dive, 2012." YouTube. Accessed August 25, 2022. https://www.youtube.com/watch?v=vYPwH54UKXo&t=1185s.
Johnsgard, Paul A. "The Niobrara as a Scenic River." In *The Niobrara: A River Running through Time*, 138–151. Lincoln: University of Nebraska Press, 2007.
Johnson, Alexander. Interview by author, June 16, 2021.
Jones, Stephen R. *The Last Prairie: A Sandhills Journal*. Lincoln: University of Nebraska Press, 2000.
Kant, Joanita. "A Historical Geography of Lake Kampeska in the City of Watertown, South Dakota." South Dakota State University Department of Geography. 2007. https://openprairie.sdstate.edu/cgi/viewcontent.cgi?article=1001&context=geo_pubs.
Koch, Emily Thornton. "The Fragrance of My Woods." In *The Outlet 2009*. Dubuque, IA: Loras College.
Koch, Kevin. *The Driftless Land: Spirit of Place in the Upper Mississippi Valley*. Cape Girardeau: Southeast Missouri State University Press, 2010.
———. "A Landing Spot for Wildlife and Humans" [Upper Mississippi National Fish and Wildlife Refuge—Savanna District]. [Dubuque] *Telegraph Herald*. December 9, 2018, 8C.
———. "Mesmerizing Mounds: Casper Bluff Full of Scenic History." [Dubuque] *Telegraph Herald*. June 8, 2014, 5E.
———. "Palisades a Treasure." [Dubuque] *Telegraph Herald*. May 29, 2011, 8C.
———. "Remembering Black Hawk." [Dubuque] *Telegraph Herald*. December 28, 2014, 5E.
Krause, David K. *The Making of a Mining District: Keweenaw Native Copper 1500–1870*. Detroit: Wayne State University Press, 1992.
Kruse Daniels, Natalie, Jen Bowman, Kelly Johnson, Dina Lopez, Amy Mackey, and Nora Sullivan. "The Legacy of Appalachian Ohio Coal Mining." In *From Surviving to Thriving in Appalachia: Place, Passion, and Possibilities*, edited by Michele Morrone and Tiffany Arnold, 55–72. Athens: The Appalachian Rural Health Institute, Ohio University, 2021.
Kruse Daniels, Natalie, PhD. Interview by author, November 17, 2021.
Landolt, Brad. Interview by author, October 20, 2021.
Larson, Douglas. *American Scientist*. January/February 2012. https://www.americanscientist.org/article/runaway-devils-lake.
LaRue, Paul. "Camp Sherman Contributes to the Destruction, and Ultimately the Preservation, of an Important Pre-Contact American Indian Earthworks." United States World War One Centennial Commission. June 4, 2019. http://www.worldwar1centennial.org/index.php/ohio-in-ww1-articles/6308-camp-sherman-versus-the-mound-city-earthworks.html.

"Last Stand of the Tallgrass Prairie." National Park Service. Tallgrass Prairie National Reserve. Accessed September 10, 2021. http://www.nps.gov/tapr/index.htm.

Least Heat-Moon, William. *PrairyErth*. New York: Houghton Mifflin, 1991.

Lensing, Melissa. Interview by author, October 7, 2021.

Lewis, Meriwether, and William Clark. "The Journals of Lewis and Clark 1804–1806." Project Gutenberg Ebook, last updated January 26, 2013. https://www.gutenberg.org/files/8419/8419-h/8419-h.htm#link12H_4_0032.

Mallom, R. Clark. "Ideology from the Earth: Effigy Mounds in the Midwest." *Archaeology* 35, no. 4 (1982): 60–64.

Mandel, Rolfe. "The Claussen Archaeological Site: Prehistory of the Flint Hills." *Symphony in the Flint Hills Field Journal* (2010). http://newprairiepress.org/sfh/2011/nature/9.

Manning, Richard. *Grassland: The History, Biology, Politics, and Promise of the American Prairie*. New York: Penguin, 1995.

Marquette, Jacques, SJ. *Father Marquette's Journal: Exploring the Mississippi River for New France, 1673–75*. Madison: Michigan Department of State, 1990.

Martin, Paul S. "Extinct Bison in Minnesota." *Ecology* 52, no. 6 (1971): 1137.

Mason, Philip P., ed. *Schoolcraft's Expedition to Lake Itasca: The Discovery of the Source of the Mississippi*. East Lansing: Michigan State University Press, 1958.

"MCI to Close Bus Plant in North Dakota." *Bus and Motor Club News*. Accessed August 25, 2022. https://www.busandmotorcoachnews.com/mci-to-close-bus-plant-in-north-dakota/.

"The Métis: A Blending of Two Cultures." Pembina State Museum. Accessed August 25, 2022. https://www.history.nd.gov/historicsites/pembina/pembinahistory4.html.

Meyers, David, Elise Meyers Walker, and Nyla Vollmer. *Carrying Coal to Columbus: Mining in the Hocking Valley*. Charleston, SC: History Press, 2017.

Mink, Claudia. *Cahokia: City of the Sun*. Collinsville, IL: Cahokia Mounds Museum Society, 1999.

"Mississippi River Facts." National Park Service. Accessed November 19, 2021. https://www.nps.gov/miss/riverfacts.htm.

"Missouri River." Encyclopedia Britannica. Accessed November 19, 2021. https://www.britannica.com/place/Missouri-River.

"Niobrara National Scenic River." National Wild and Scenic Rivers System. Accessed August 26, 2021. https://www.rivers.gov/rivers/niobrara.php.

"Niobrara Valley Preserve." Nature Conservancy. Accessed August 26, 2021. https://www.nature.org/en-us/get-involved/how-to-help/places-we-protect/niobrara-valley-preserve/.

Norris, Kathleen. *Dakota: A Spiritual Geography*. New York: Houghton Mifflin, 1993.

"Orangeville Rise, Indiana." Nature Conservancy. Accessed September 29, 2021. http://www.nature.org/en-us/get-involved/how-to-help/places-we-protect/orangeville-rise.

"The Ore Docks in Marquette." Travel Marquette. Accessed August 25, 2022. https://www.travelmarquette.com/things-to-do/arts-culture/ore-docks/#:~:text=The%20Iron%20Ore%20Dock%20in,by%20the%20Cliffs%20Natural%20Resources.

"Our History." Lincoln University. Accessed November 22, 2021. https://www.lincolnu.edu/web/about-lincoln/our-history.

Parks, Soshi. "York Explored the West with Lewis and Clark, but His Freedom Wouldn't Come until Decades Later." *Smithsonian Magazine*. March 8, 2018. https://www.smithsonianmag.com/history/york-explored-west-lewis-and-clark-his-freedom-wouldnt-come-until-decades-later-180968427/.

"Passion and Persistence Pays Off for Dakota Elder." US Fish and Wildlife Service. November 24, 2021. https://usfws.medium.com/passion-and-persistence-pays-off-for-dakota-elder-2ac15cb923f9.

Patterson, Eric. Interview by author, August 6, 2021.

"Phase-Out / Abandonment of the SRMSC." SRMSC. Accessed August 25, 2022. https://srmsc.org/int2070.html.

Pike, Zebulon Montgomery. *Exploratory Travels through the Western Territories of North America*. London: Longman, Hurst, Rees, Orme, and Brown, 1811.

Powell, Wayne. "Indiana Limestone." City University of New York. Accessed September 29, 2021. http://academic.brooklyn.cuny.edu/geology/powell/613webpage/NYCbuilding/IndianaLimestone/IndianaLimestone.htm.

Roeder, James A. "Niobrara National Scenic River, 1985–2000: Old Arguments, New Compromises." University of Nebraska-Omaha. September 1, 2002. Digital Commons@UNO. https://digitalcommons.unomaha.edu/cgi/viewcontent.cgi?article=1582&context=studentwork.

Rogers, Ann. *Lewis and Clark in Missouri*. 3rd ed. Columbia: University of Missouri Press, 2002.

Rooks, David. "This Date in Native History: Massacre at Whitestone Hill." *Indian Country Today*. September 13, 2018. https://indiancountrytoday.com/archive/date-native-history-massacre-whitestone-hill.

Rooks, Hannah. Interview by author, August 1, 2022.

Sandoz, Mari. *Old Jules*. Lincoln: University of Nebraska Press, 1935.

Santa Fe Historic Trail. Signboard, National Park Service. City of Council Grove. July 23, 2021.

Schoolcraft, Henry Rowe. "Schoolcraft's Narrative of an Expedition through the Upper Mississippi, to Itasca Lake." In *Schoolcraft's Expedition to Lake Itasca*, edited by Philip P. Mason. East Lansing: Michigan State University Press, 1958.

*The Search for the Great River's Source*, booklet. Park Rapids, MN: Itasca State Park, n.d.

*Sentinel Lake Assessment Report Elk Lake, Clearwater County, Minnesota*. Saint Paul: Minnesota Department of Natural Resources, 2011.

Severin, Timothy. *Explorers of the Mississippi*. New York: Knopf, 1967.

Shaffer, Barb. Interview by author, October 6, 2020.

Sheldrake, Philip. *Spaces for the Sacred: Place, Memory, and Identity*. Baltimore: Johns Hopkins University Press, 2001.

"Shutdown of ABM Facility Threatens Future of North Dakota Town." *The Dispatch*. December 8, 1975, 10.

Sieber, Ellen, and Cheryl Ann Munson. *Looking at History: Indiana's Hoosier National Forest Region, 1600–1950*. Bloomington: Indiana University Press, 1994.

Sikes, Robert S., John R. Tester, and Ben Thoma. *Mammals of Itasca State Park*. Saint Paul: Minnesota Department of Natural Resources, 2003.

"Sleeping Bear Dunes National Lakeshore History." OhRanger. Accessed August 25, 2022. http://www.ohranger.com/sleeping-bear-dunes/history.

Smith, Andrew, and Rachel Klabacka. "Archaeological Investigations in the Upper Wabash River Valley: A 2009 Survey in Huntington, Miami and Wabash Counties, Indiana." http://bsu.edu/-/media/www/departmentalcontent/aal/aalpdfs/roi%2076-100/roi%2076%20vol%201%20public.pdf?la=en&hash=CA44527B359FD00CB91A4C4456196C304F7EAA3A.

Smith, Julian. "Righting Wrongs at Effigy Mounds." *American Archaeology* 23 (Spring 2019): 12–18.

"Soil Survey of Pembina County, North Dakota." United States Department of Agriculture Soil Conservation Service. July 1997.

South Dakota State University Museum of Agriculture. June 28, 2022.

Sprenger, Matt. Interview by author, July 28, 2021.

"SRMSC Construction/Engineering." RSL3 Missile Site Tours. Accessed August 25, 2022. https://rsl3.com/site-tours.html.

Stahl, Frank. *1865 Santa Fe Trail Diary of Frank Stahl.* Accessed February 28, 2022. http://www.frankstahlbio.net/trail_diary_web.htm.

"State-by-State Visualizations of Key Demographic Trends from the 2020 Census." United States Census Bureau. Accessed December 9, 2021. https://www.census.gov/library/stories/state-by-state.html.

"State of Ohio Invests $2 Million to Build Out One of the Largest Mountain Bike Trail Systems in the US." *Highland County Press.* July 2, 2021. http://highlandcountypress.com/Content/Sports/Sports/Article/State-of-Ohio-invests-2M-to-build-out-one-of-the-largest-mountain-bike-trail-systems-in-the-US/3/21/69789.

Steck, Joe. "Logging Boom and the Birthplace of Minnesota." *Mankato Times.* March 14, 2017. http://mankatotimes.com/2017/03/14/a-moment-in-time-logging-boom-and-the-birthplace-of-minnesota/.

Tanner, Beccy. "Kanza People Returning to Sacred Land near Council Grove." *Wichita Eagle.* April 22, 2015. http://www.kansas.com/news/state/article19178235.html.

Trask, Kerry. *Black Hawk: The Battle for the Heart of America.* New York: Henry Holt., 2006.

Treinen, Paul. Interview by the author, July 20, 2022.

Tvedten, Benet, OSB. "The View from a Monastery." In *Inheriting the Land: Contemporary Voices from the Midwest,* edited by Mark Vinz and Thom Tammaro, 309–312. Minneapolis: University of Minnesota Press, 1993.

"The Two Hundred Largest Cities in the United States by Population, 2021." World Population Review. Accessed December 9, 2021. http://worldpopulationreview.com/us-cities.

Tyrell, Joseph. *David Thompson, Explorer.* n.p., 1900, digitized by University of Toronto. https:/archive.org/stream/davidthompsonexpootyrr/davidthompson expootyrr_djvu.txt.

Tyrrell, Kelly April. "As the River Rises: Cahokia's Emergence and Decline Linked to Mississippi River Flooding." *University of Wisconsin-Madison News.* May 4, 2015. https://news.wisc.edu/as-the-river-rises-cahokias-emergence-and-decline-linked-to-mississippi-river-flooding/.

"Underground Railroad: Vesuvius Forest." United States Department of Agriculture: Forest Service. Accessed October 14, 2021. http://www.fs.usda.gov/recarea/wayne/recarea/?recid=81885.

"Unemployment Rate in Cavalier County, ND." FRED Economic Data. Accessed August 25, 2022. https://fred.stlouisfed.org/series/NDCAVA9URN.
"Upper Mississippi River National Wildlife and Fish Refuge." US Fish and Wildlife Service. Accessed June 17, 2016. www.fws.gov/refuge/upper_mississippi_river.
"Valentine National Wildlife Refuge." US Fish and Wildlife Service. Accessed August 26, 2021. https://www.fws.gov/refuge/valentine/.
"Volunteers Will Drive Longhorn Cattle 180 Miles to Fort Robinson." *Yankton Daily Press and Dakotan*. November 4, 2000. https://www.yankton.net/news/article_d1b68b1f-6baf-551e-b15d-6314db3fa114.html.
Wahl, Scott. Interview by the author, July 25, 2022.
Walker, Diane. "Lick Creek Settlement Holds Piece of Black History in Indiana." *Limestone Post*. February 27, 2019. http://limestonepostmagazine.com/lick-creek-settlement-holds-piece-of-black-history-indiana/.
Wilkie, William E. *Dubuque on the Mississippi*. Dubuque, IA: Loras College Press, 1987.
Wilson, Jennifer. "The Park That Wasn't." *Iowa Outdoors* 72, no. 1 (January/February 2013): 38–44.
Wisconsin Department of Natural Resources. *The Ecological Landscapes of Wisconsin*. Madison: Wisconsin Department of Natural Resources, 2015.
Wollmer, Michael. Interview by the author, January 21, 2021.
"Yankton Sioux Treaty Monument." National Park Service. Accessed August 25, 2022. https://www.nps.gov/mnrr/learn/historyculture/yankton-sioux-treaty-monument.htm#:~:text=This%20treaty%20provided%20for%20the,signed%20for%20the%20federal%20government.

**KEVIN J. KOCH** is Professor of English at Loras College in Dubuque, Iowa. He is author of *Skiing at Midnight: A Nature Journal from Dubuque County, Iowa*; *The Driftless Land: Spirit of Place in the Upper Mississippi Valley*; and *The Thin Places: A Celtic Landscape from Ireland to the Driftless*. More of his work, including shorter works on outdoor places and photos, can be viewed at https://www.kevinkochdriftlessland.net/.

*For Indiana University Press*

Tony Brewer *Artist and Book Designer*
Dan Crissman *Trade and Regional Acquisitions Editor*
Emma Getz *Editorial Assistant*
Samantha Heffner *Trade Acquisitions Assistant*
Brenna Hosman *Production Coordinator*
Katie Huggins *Production Manager*
David Miller *Lead Project Manager/Editor*
Dan Pyle *Online Publishing Manager*
Pamela Rude *Senior Artist and Book Designer*
Stephen Williams *Marketing and Publicity Manager*